Prepping and Homesteading

What You Need to Know to Be Self-Reliant When STHF, Including Tips on Stockpiling, Growing Your Own Food, and Living Off the Grid

Contents

PART 1: PREPPING 1

INTRODUCTION 2

PART ONE: ESSENTIAL PREPPING 4

UNDERSTANDING PREPPING 5

6 MISCONCEPTIONS ABOUT PREPPERS 7

15 BENEFITS OF PREPPING 9

21 THINGS TO KNOW BEFORE BECOMING A PREPPER 13

THE BEGINNER'S PREPPING TOOLKIT 18

PREPPERS TOOLKIT CHECKLIST 28

CONSCIOUS STOCKPILING: ITEMS YOU SHOULD NOT SPEND
YOUR MONEY ON 34

31 ESSENTIAL PREPPING SKILLS TO KNOW 39

TOP 15 ROOKIE PREPPER MISTAKES TO AVOID 48

PART TWO: OFF-GRID LIVING 55

LIVING OFF THE GRID: REASONS AND MISCONCEPTIONS 56

COMMON MISCONCEPTIONS ABOUT OFF-GRID LIVING 60

THE REALITIES OF LIVING OFF-GRID 64

HOMESTEADING 101 68

DIFFERENT TYPES OF HOMESTEADING 68

BASIC HOMESTEAD STEPS 70

SOLAR ENERGY AND OTHER POWER OPTIONS 73

WATER SOURCES, SOLUTIONS, AND SYSTEMS............77

OFF-GRID WATER SOURCES..78

OFF-GRID WATER UTILIZATION SYSTEMS 80

DETERMINING WHEN WATER DOESN'T NEED FILTERING 82

THE OFF-GRID BUDGET: HOW MUCH WILL IT COST? 83

MAINTENANCE COSTS 88

PART THREE: SHTF SURVIVAL..90

10 SHTF SCENARIOS: WHAT TO EXPECT AND WHAT TO DO 91

SHTF EVACUATION 101

MEDICAL CARE DURING SHTF...106

ADDITIONAL FIRST AID CHECKLIST..................................114

THE BUG-OUT BAG: SURVIVING SHTF ON THE GO 116

WILDERNESS SURVIVAL TIPS..120

CONCLUSION 124

PART 2: HOMESTEADING 125

INTRODUCTION...126

PART ONE: HOMESTEADING BASICS 128

HOMESTEADING EXPLAINED..129

WHAT TYPE OF HOMESTEADER ARE YOU? 134

THE URBAN/APARTMENT HOMESTEADER 135

THE MEDIUM TO LARGE-SCALE HOMESTEADERS 137

STEPS BEFORE YOU START HOMESTEADING 140

11 ESSENTIAL HOMESTEADING SKILLS 147

HOW MUCH DOES HOMESTEADING COST?......................156

HOW TO AVOID THE BIGGEST HOMESTEADING MISTAKES 162

THE MOST COMMON HOMESTEADING ERRORS.............................162

PART TWO: LIVING OFF THE LAND 167

FINDING LAND FOR YOUR HOMESTEAD................................168

OTHER OPPORTUNITIES...174

GROWING YOUR OWN FOOD 175

FRUITS AND VEGETABLES 181

MEDICINAL HERBS 185

21 HOMESTEAD GARDENING HACKS 187

SELF-SUFFICIENT FARMING: WHAT LIVESTOCK AND HOW MANY?..194

ANIMALS YOU CAN RAISE ON YOUR LARGE HOMESTEAD 195

HOW TO RAISE BACKYARD CHICKENS 205

YOUR HOMESTEAD PANTRY..217

THE ESSENTIALS 218

DECLUTTER, THEN ORGANIZE 222

PART THREE: HOMESTEADING SURVIVAL 223

EXTREME WEATHER CONDITIONS ON THE HOMESTEAD 224

LIVING WITH HOMESTEAD PETS 229

HOMESTEADING ALONE...234

GAME-CHANGING DECISIONS 235

MONEY-SAVING CRAFTS AND DIY PROJECTS 239

DIY BACKYARD INCINERATOR 241

DIY CITRUS ALL-PURPOSE CLEANER 242

DIY CANDLES...244

DIY DETERGENT 245

DIY CURTAINS...246

DIY BASKETS 247

DIY Paint 247

14 HOMESTEADING INCOME IDEAS 249

CONCLUSION 255

REFERENCES 257

Part 1: Prepping

An Essential Survival Guide for DIY Preppers Who Want to Be Self-Reliant When SHTF, Including Tips for Living Off the Grid, Homesteading, and Stockpiling Properly

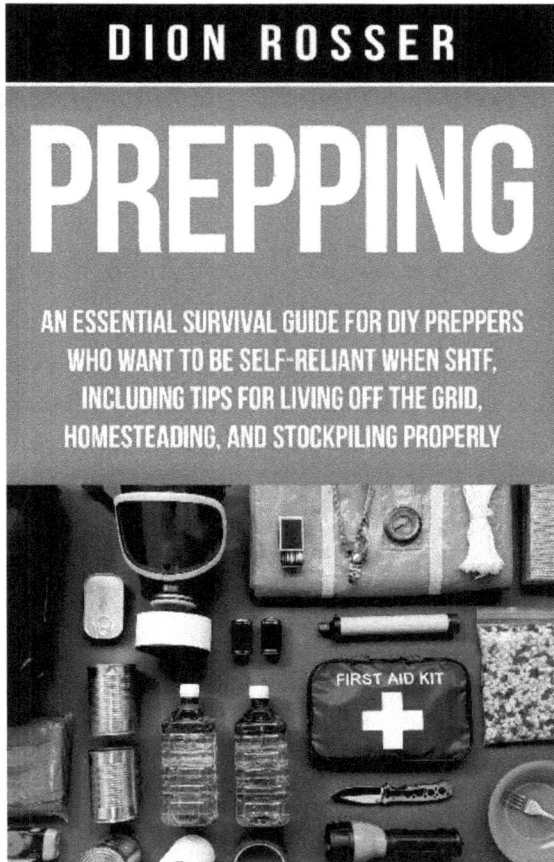

Introduction

Thank you for choosing *Prepping: An Essential Survival Guide for DIY Preppers Who Want to Be Self-Reliant When SHTF, Including Tips for Living Off the Grid, Homesteading, and Stockpiling Properly* as your preppers guide.

Obviously, you have an interest in learning how to be prepared, and congratulations on taking the right first step.

Prepping is all about being prepared for the worst-case scenario. To begin, look at the example of Super Storm Sandy. Sure, people knew it was coming; they had more than enough warning. They even knew what path it was headed for, and where it would strike first. Were they prepared? No way! Most people affected had lived through bad storms in the past, and that is what they based their prepping on. They didn't prepare for the *absolute worst-case scenario*, and that is what they got.

Most people didn't fully understand how the storm surge would affect them, how the flooding that followed would wreak havoc on their homes and lives. They followed the advice given – to stockpile 72 hours' worth of supplies –, but it wasn't enough. What officials didn't take into account when they drew up their guidelines was the infrastructure – how much it had aged and deteriorated.

And that is why you are here because you don't want to be caught out unprepared. That is what this guide is for – to give you the most up-to-date information to make sure you can survive. And face it – the way the world is going now, anything can happen. The weather is getting more extreme; wars are springing up everywhere, countries like Australia, states like California, even the Amazon, are burning, destroying precious resources and cutting people off, leaving them with no hope of rescue for some time. 72 hours' worth of supplies won't do them any good, and it won't do you any good either.

This book has been written in a way that you will easily understand what to do. It includes step-by-step guides, hands-on tutorials, expert advice, and so much more. There is no fluff, no out-of-date information, and no hard to follow advice. Everything here is simple, sensible, and if you follow it properly, it is pretty easy to implement.

In no time at all, you will be a pro prepper, so dive in now and start this journey of discovery!

PART ONE: ESSENTIAL PREPPING

Understanding Prepping

At a fundamental level, prepping is a shortened version of the words "preparation" or "preparing", but its modern use has taken it much further. Now, the word is associated with planning and prepping for disaster scenarios and major catastrophes. It involves basic things, such as stockpiling food, water, medicines, and so on – yet it goes much deeper than that.

To some people, the very idea of stocking up for a large-scale disaster is nothing short of stupid, but the last few years should have taught people differently. The number of weather and war-related disasters is growing, becoming more commonplace than ever before, and it seems now that prepping is the obvious way to go. Who knows when the next disastrous flood will hit? The next out-of-control bushfires? Tornadoes and hurricanes? And, God forbid, who knows when and who will hit that red button?

Being unprepared is to be caught out big style; those who take the time to start prepping now will have a much better chance of survival than those who go through life thinking it will never happen.

Ask yourself this question – *do I have any kind of insurance? Health, life, home?* If you do, that makes you a prepper. Insurance is your preparation for when something goes wrong, and the prepping being talked about here is the same thing. By building up your stockpiles and learning new skills, you are insuring yourself and your family against a future disaster that could lead to society collapsing.

Do you watch the news, read the papers, or listen to the radio? If you do, that means you are well aware of the rise in human-made and natural disasters. Just because they have not happened to you yet doesn't mean they won't, and there is no better way to be ready than to start prepping right now.

The Cold War with Russia was not that long ago, and people still live under the threat of nuclear war every day. Face it – the United States of America doesn't exactly have the best relations with some countries, does it? What about terrorist attacks? Natural disasters? And what if another global pandemic, like the Spanish flu, struck? It could kill billions of people, and the impact on life would be horrendous, not to mention how the economy and society would be affected. How?

Electricity supplies could fail, stores would likely close, and food and water supplies could be heavily depleted. In short, the world as you know it now is in trouble, more than ever before, and everything you have come to take for granted could be gone in seconds. It could happen tomorrow, it could happen next year, in 200 years, or never, and that is the point of prepping – nobody knows when the next disaster will hit; it is just that the preppers are ready for it.

6 Misconceptions about Preppers

These are six of the most common misconceptions about preppers:

1. Preppers are paranoid conspiracy theorists.

Television can take the blame for this. Many people think preppers are "weirdos", paranoid ones at that. The truth is very different. Preppers are people who simply prefer not to rely on others for their survival should the SHTF. They are actually calm and balanced people who are ensuring they are prepared for anything.

2. Preppers are extremists who believe the apocalypse is about to hit.

Not at all. Prepping is nothing to do with zombies or government collapses. It is to do with being prepared for survival in any situation.

3. Preppers are isolated.

Sure, some may live out in the wilderness, but your next-door neighbor could be a prepper for all you know. It does not matter where you live; you can learn how to be self-sufficient, build fires, and purify water.

4. Preppers are gun fanatics.

It could seem a little extreme when a prepper becomes familiar with self-defense and using guns; however, that doesn't make them a gun fanatic. It is not a bad thing to learn how to protect yourself, so long as your firearms are kept for that reason alone.

5. Preppers live in bunkers and are constantly prepping.

Some probably do, but it is not the norm. Preppers are ordinary people, living ordinary lives; it may be a way of life, but it doesn't take over their whole life.

6. Preppers are rich.

After all, prepping costs a lot of money, doesn't it? It could do, but it doesn't have to be. You can start prepping now with just a few dollars each month. Most people build up slowly, buying extras here and there. It does not have to cost a fortune and, in many cases, will barely make a difference to the budget you have now.

15 Benefits of Prepping

While there are tons of reasons why everyone should start prepping, so many feel it is a waste of their time. So, before you begin prepping or dismiss it out of hand completely, you need to understand the immediate benefits it brings. And these go beyond the main benefit – a better chance of survival during disaster scenarios. So, if you needed any more encouragement to start prepping right now, here are 15 excellent reasons why:

1. Save Money

Survival doesn't need to cost a fortune; there are loads of things you can do right now to save money, including:

- Growing your own food
- Making your home safer without forking out for costly alarms
- Make your fitness equipment
- Learn to DIY around the house
- Stock up on dried and canned foods a little at a time

2. Better Health

Doing drills, going for hikes and eating better, organic food all lead to a healthier you – being prepared for when the SHTF means getting in shape now. You'll be eating less junk food and consuming more vitamins, minerals, and macro-nutrients to help keep your body healthy too.

3. Better Relationships

You've got no choice – planning for doomsday means working together, and prepping can pull even the most broken families back together again. Going for a hike, a camping trip, and even watching survival shows on TV can all help.

4. You Become Self-Reliant

Reliance on others is too prevalent in today's society, and although it won't be possible to go entirely off-grid, you can learn self-reliance in many ways – how far you take it is up to you.

5. Leadership Skills

If you want to lead your family in survival, you need to become a leader. You must have a vision, learn to encourage others, solve conflicts, and be the leader you need to be. However, you need to learn how to be a leader without coming across overly critical and bossy!

6. Learn Responsibility

Becoming a prepper means being responsible, not just for yourself but also for those in your prepper family.

7. You Could Make Money

By selling surplus crops to the neighborhood or at the market (by using your survival skills), you learn to help others while earning an income, and even teach others how to survive.

Just be aware of the tax laws surrounding earning extra money and requirements for handling food.

8. Never Run Out of Loo Paper Again!

You may be laughing, and it may not be the best reason to begin prepping, but it is true!

9. You Won't be so Stressed

Instead of spending your life complaining about what the world is coming to, spend it preparing instead. There won't be such a weight on your shoulders, and you can go through life feeling happier, knowing that, whatever happens, you are prepared for it.

10. New Hobbies

Camping, hiking, fishing, learning to build fires, learning to find water sources – the list is endless. Add to that growing your own food, learning new cooking skills, woodwork, DIY, and so on.

11. You Could Save a Life

Learning basic first aid and survival medicine is always handy, and you could, one day, use your skills to save a life when no other help is there.

12. You won't be So Bored or Lonely

Many people have nothing to do and see no one. Prepping changes all that; not only are you kept busy with your new skills, but you can also join a preppers group and meet new people.

13. You Learn Appreciation

Everyone is guilty of forgetting the importance of the small stuff – a bottle of water, an apple from your tree, completing that ten-mile hike. Count your blessings!

14. Rediscovering Nature

Do you spend your day divided between the office and your couch at home? Going hiking and camping gets you out there in nature, and that will have positive benefits for you and your life.

15. Learning to Negotiate

When you become a prepper, you learn to barter, and that is a form of negotiation. You don't need to wait until the last minute – start bartering now and refine your skills.

Now you know why you should start prepping, read on to learn some other things you need to know.

21 Things to Know Before Becoming a Prepper

Most people make mistakes when they start prepping. For most, it is because they rush in, not taking the time to think and plan ahead; for others, it's a lack of decent information. So, here is a list of 21 things you need to know so you do not make the same mistakes, and waste time, energy and money when you don't need to.

1. Live Below Your Means – Now

Forget having to use a credit card to purchase all your prepping supplies in one hit – find ways to bring your bills down now and put the money you save to one side.

2. Don't Spend Every Penny in Month One

Prepping is a slow and steady business. You might think you need to get that survival item now, but you should look around for deals and wait – patience pays off when you find a better price, or you already have a substitute that will do just as well.

3. Start Storing Water

You need water far more than you do food, and a few liters won't get you anywhere. Start storing it now. You don't have to go out and spend a fortune on bottled water; collect your own and store it in clean barrels and collapsible containers in your basement or garage – stored right, it can be kept for long periods.

4. Don't Use Old Milk Jugs to Store Water

It might seem like a good idea, but it's doomed from the start. You can never get all the milk residue out of these jugs, and that can lead to your water becoming home to harmful bacteria. Plus, being plastic, the jug will break down over time, and that's a mess you don't want.

5. Don't Waste Money on Food That Won't Get Eaten

You might think you found a great deal on a boatload of canned spinach, but does anyone eat it? It's just a waste of precious resources – time, money, and storage space. By all means, look for great deals, but if it doesn't get eaten, don't waste time and money on it.

6. Don't Focus Only on Canned Foods

Some people have a strange idea that they should only store canned goods. Wrong – you need a decent variety of foods, including canned, freeze-dried, and dry to ensure you have a decent diet. Otherwise, you run the risk of food boredom setting in. Not only that, canned foods are loaded with sodium, and too much isn't a good thing.

7. Make Sure Your Storage Shelves are Sturdy

You might think you got a great deal on those particleboard shelves, but once you start piling goods onto them, they won't last long. Use wire shelving or build your own sturdy wood or metal shelving.

8. Don't Store Everything in One Place

One badly timed disaster could wipe out your entire supply of food and other supplies. Keep your food and water separated into several caches, make sure you have a bug-out bag on hand at all times (and one in each vehicle), and keep a few supplies at your bug-out shelter location.

9. Prepping is About More Than How Much You Store

It's about learning skills and having the right knowledge to help you survive. You'll need training for some and lots of hands-on experience – at the end of this section, you will find a list of those skills.

10. Hygiene and Sanitation Are Important Too

Storing food and water is one thing, but don't forget soap and toilet paper too. Cleanliness is important in avoiding illness – the last thing you want in a survival situation is to fall ill, especially when hospitals will be overrun or, worse, closed altogether.

11. Don't Forget Special Needs

There may be people who need wheelchairs, oxygen, insulin, and so on; make sure you include them in your prepping.

12. Got Pets? Don't Forget Them

When it comes to a survival situation, you have two choices with pets – abandon them or care for them. Most people will choose the latter, but make sure you make the decision now. Then start stockpiling food and other necessary items for them – and don't forget to factor them into your water calculations.

13. Get Family Members On Board

Everyone needs to be in on this; they don't need the same excitement level as you, but they do need to know how to do things and have a certain amount of knowledge. Get them involved from the start.

14. Don't Broadcast Your Plans to Everyone

You don't need everyone knowing what you are doing and what you have in stock. If the SHTF, you don't want the whole neighborhood at your door; you can't help your own family and everyone else too. Keep your plans between you, your family, and a few trusted friends.

15. Keep in Shape

If you don't, the first day of any disaster scenario is going to leave you exhausted. You'll be hiking about, taking supplies to and from, repairing damage, and it will leave you dead on your feet. It's easier than you think to stay in shape – a power walk for 30 minutes every day will do the trick.

16. Don't Assume Guns and Ammo Can Keep You Safe

Yes, you should have guns and plenty of ammo to defend yourself and your family, but you should avoid confrontation where possible. Learn stealth and learn not to draw attention; guns can't keep you safe from others who have them.

17. Make a Plan to Get Home

Many people forget that disasters don't wait for them. They can happen when you are anywhere, so make sure you have a plan in place to return home or a safe meeting place for you and your family.

18. Never Make Assumptions

Some think that they will need to bug out, and others plan to bug in. The thing is, you have no idea of knowing what's going to happen, so have a plan in place and then have a backup plan. And another if you need it.

19. Test Your Tools

Never assume that a tool is going to work when you really need it – test it out. And don't stockpile a dozen of the same tool – if it doesn't work, you may need a different one. Not only that, the more tools you stockpile, the more you have to carry.

20. Small Steps

Many DIY projects need a lot of time and education to complete, and patience is the key. Don't rush; take your time, or you will end up frustrated and tired. Many small steps go a long way.

21. The World Isn't Going to End Tomorrow

Well, it might, but there's a good chance you'll get a little warning. The problem is some preppers get into the habit of thinking that way and panic; that leads to bad decisions. Always be prepared, but don't forget to enjoy life on the way. Don't lose yourself so much in your doomsday prepping that everything else passes you by and always keep one thing in mind – there is a chance that doomsday won't happen.

The Beginner's Prepping Toolkit

When it comes to prepping and building up an essential toolkit, there are many things to consider. These are the top ten items and, although most people say you should have enough for 72 hours, you should really be planning for a minimum of two to four weeks.

Water

There is a reason this is at the top of the list; as a rule, you can go for at least three weeks eating little to nothing and still survive, but water is a different matter. Depending on conditions, you cannot go without water for more than three or four days, and that is being generous. So, what should you plan for?

You need to store at least one gallon per person per day. If you are prepping for four, aim for a minimum of 28 gallons per week (56 for two weeks and 112 for four). But do keep in mind that you don't just need water for drinking; you need it for cooking and washing too and, if you have pets, factor those into your calculations as well. The best thing to do is to aim for two gallons per day.

You can start by purchasing water at your grocery store, but thinking long term, you should acquire large plastic water storage containers – a mixture of small and large. You will also require water purification tablets and a portable water filter, in case you need to forage for water outside. This is important – drinking dirty water can introduce bacteria and harmful pathogens, and the last thing you need is to be sick on top of trying to survive. You may also use regular, non-scented chlorine bleach, at a rate of two drops per liter, to purify water.

Food

The next most important item is food, and again, aim for two to four weeks. You should be starting to stockpile canned foods – soups, meats, fruit, and vegetables, but do make sure they have a decent shelf life and are not dented. You should also be looking at storing dried foods, such as rice, pasta, oatmeal, flour, lentils, and beans. Try to store these in airtight containers, especially the flour, as it can go bad if not stored correctly, and may also attract unwanted rodent and insect attention.

Think about storing sugar, salt, olive or coconut oil, canned cheese and butter, powdered eggs and powdered milk, along with tea and coffee. You can purchase MREs online (Meals Ready-to-Eat), usually from military surplus stores, along with dehydrated and freeze-dried foods. An idea of a list for two weeks' storage is:

- 20 lbs. of beans
- 20 lbs. of rice
- 20 cans of fruit
- 20 cans of vegetables
- 20 cans of meat
- Two large containers of peanut butter
- Two large bags of flour
- One bag of sugar
- One bag of salt*
- One pound of oats
- One gallon of olive or coconut oil

*Don't forget that you can use salt as a preservative for meat and fish, so stockpile as much as you can now.

Obviously, you would only be storing the kinds of foods you and your family eat, and you can change this list to suit your preferences.

Try to aim for a reasonable balance of carbohydrates, fats, and proteins and keep a supply of multivitamins on hand too.

First Aid

People take emergency services personnel for granted these days, so much so that most people are unprepared in terms of their own first aid supplies. Should the SHTF, you will need first aid supplies, and there are two ways you can go about it – a basic first aid kit or a full-on medical trauma kit.

At the very least, you and all members of your preppers group should take a first aid course. You should also purchase a survival medicine guide and familiarize yourself with it. If you want, you can purchase survival first aid kits online, or you can make your own. In the checklist at the end of this chapter, a full list of what you need is provided, but as a guide, you will need bandages, Band-Aids, sterile gloves, tweezers, antibiotic and antibacterial creams, and over-the-counter medications.

If you are on prescription medications, you must see your doctor and ensure that you have a decent supply, just in case, and ensure that you have plenty of asthma inhalers on hand if you are asthmatic.

Don't forget; the emergency services are going to be tied up, focusing their efforts on major-impact areas and are not likely to get to you straight away, if at all.

Sanitation

Correct sanitation is incredibly important in an SHTF scenario, more so if the electricity and water are out of action. Without it, diseases will soon spread, and that's the last thing you need in this situation.

There are several ways to go here. If your property is on a septic system, usually those properties in outlying or country areas, then you can use your toilet as normal. If there is no running water, you can fill the cistern manually.

If you are not on a septic system, the first thing to do is ensure that the mains sewerage is working – if not, do NOT flush your toilet under any circumstances. Do so, and you risk the sewerage backing up and coming up through your water lines, bathtub, basin, and so on.

In a situation where there is no running water, or you have decamped to your bug-out station, there are a couple of things you can do. A five-gallon bucket double-lined with heavy-duty trash bags may be used as a toilet – you can pop a toilet seat on top if you want. After each use, a generous handful of cat litter or dirt with a little disinfectant spray should be layered on top. When the bag is two-thirds full, cover with litter or dirt and tie it up. Store it out of the way in a sealable container or take it a long way off from where you are living (at least 200 feet), dig a hole, and empty the bag contents into it.

Alternatively, if it is a short-term solution you need, you can dig catholes, again, away from your living area, and make sure they are well-filled in afterward.

Do make sure you have plenty of antibacterial soap, hand sanitizer, and biodegradable toilet paper on hand, as well as a decent supply of wet wipes.

Cooking

It's one thing making sure you have a decent supply of food on hand, but, in many cases, you will need a way of cooking it. In most disaster scenarios, the first thing that goes is the electricity supply, and there is every chance it could be off for weeks, even months. Along with that, natural gas supplies may be cut off too.

What you need depends on whether you are bugging in or out. Bug in, and you can go with a propane BBQ (you probably have one in your backyard) or a portable gas stove – for both, you will need a supply of fuel on hand. You could also purchase a supply of disposable BBQs for a quick meal.

If you are bugging out, you can get away with a portable gas stove or building a fire. In both cases, you need quality, durable cooking utensils – you already have these in your home, but do you really want to lug a cast-iron skillet around with you if you have to bug out?

Look for stainless steel pots and pans, tin or stainless steel plates (you can also go for disposable) and quality knives, forks, and spoons. You can purchase kits that take care of all of this, or you can make your own. You must make sure you have a manual can opener, a bottle opener, a decent supply of matches, tin foil, and plastic bags for disposing of leftovers and rubbish.

And don't buy the first or cheapest items you see – do your homework, read reviews, and be sensible about things.

Power

The first thing you need is a good flashlight, one for every person in your group. Military or police-grade flashlights are a good option because they have stronger beams, and some have the option of SOS flashing on them. Alongside that is a good supply of batteries, enough for several weeks. Alternatively, purchase rechargeable batteries. You should also keep a couple of wind-up torches on hand as well; they are not that strong, but they don't require batteries and will do in an emergency.

If you are bugging in, then a portable generator is the way to go – you can get a gas-powered one, which means keeping a supply of gasoline on hand, or you can go for a solar-powered generator. You can also purchase battery banks, which can help with short-term charging.

Cash

A disaster scenario is when cash truly becomes king – if you have it, you can get almost anything. Start stashing away small bills – aim for about $1,000 as a minimum, although this will depend on how many people are in your survival group. The reason why having small bills is two-fold – first, change is unlikely to be readily available, and second, if you do need to barter, you don't want to be using large notes, and potentially overpaying for things. Plus, it doesn't hurt to let people think you only have a few dollars spare.

Keep your cash reserves safe. First off, don't store it all in the same place. Split it into equal amounts and hide it in different locations – keeping it well hidden, out of sight. Some people choose to bury their cash, wrapping it in several plastic bags first, to keep moisture out. This is fine, so long as you remember where you buried it!

Be aware that, in emergencies, homes are the perfect target for looters – keeping your cash somewhere obvious is a big no-no. Get creative; just make sure you know where it is.

Communications

Communication is an important factor in any SHTF scenario. The phone lines are likely to be out of commission, and there is little chance of your mobile phone having any coverage either. You also need to be in touch with others in similar situations and to hear of any news.

What you need, at a minimum, is a two-way radio, preferably one per person in your group or ham radios, and a way of keeping them powered. Typically, these run on batteries, so make sure you have a decent supply. You should also have a radio of some description, a battery-powered or wind-up radio. That way, you can keep up to date with any reports coming in on the situation at hand and what, if any, emergency help is on the way.

It is worth noting that, although it is illegal to broadcast on ham radios, it has been deemed legal in the case of an emergency.

Mobility

Mobility is also an important consideration. If you are in a position to leave your home and evacuate somewhere safe, then you should ensure that you have sufficient fuel to get you away. If the situation is such that traveling simply isn't an option, you have to decide whether to bug in or bug out. In most cases, bugging in won't be an option –

electricity, water, and gas lines will be out of commission, and if the disaster is such that your home has been damaged, you will need to get out.

In that case, you need your bug-out bag, and that, as a minimum, should contain emergency shelter, water, food, and security. In part three of this guide, there is more detail on the bugout bag, but for now, here is an idea of what you should have in it – and each person should have one:

- Tent
- Sleeping bag
- Space blanket
- Water bottle or bladder
- Portable filter
- Food rations
- Gloves
- Jacket
- One change of clothes
- Warm headgear
- Matches
- Flashlight
- Headlamp
- Basic first aid kit
- Map of the area
- Compass
- Mini shovel
- Ax/hatchet
- Paracord
- Multi-tool
- Knife
- Pepper spray
- Charger – solar or battery

- Whistle
- Goggles
- Copies of your important documents
- Passport
- Titles and contracts
- Address book
- Family disaster plan
- At least $500 cash – small bills
- Prescription medication
- Small mirror

Self-Defense

Bugging in or bugging out, self-defense is vital. In disaster scenarios, anything can happen, and you need to bear in mind that most people will only be prepared for a couple of days – after all, FEMA (Federal Emergency Management Agency) recommends you prepare for just 72 hours. Everyone knows that most disaster situations continue long after this, and so long-term planning is crucial.

When people get desperate, they will do anything to survive, and if that means attacking you to get what you have, they will. And they won't think twice about hurting you either. If you are bugging in, there are certain things you can do to ensure a certain level of safety – make sure you have high fencing all around the property, set up obstacles to slow down would-be attackers, make sure all your windows and doors are bolted shut, and so on. However, in the event that someone makes it past all that, or you are bugging out, you need to consider how you will defend yourself.

You should keep weapons on hand. At the very least, have stun guns and/or Tasers. Stun guns require that you have direct contact with your attacker, while Tasers can be fired from a distance – while most Tasers don't have a strong enough shock capacity to need a permit, you must check regulations in your area. Having said that, in a real apocalypse-type scenario, not many people will be worried about regulations!

Pepper spray is one of the best defense weapons – it is easy to conceal, light, and simple to use. Knives are easily concealed, but again, unless you are a professional knife thrower, you need to be up close and personal with your attacker.

And then there are guns. Not everyone is comfortable carrying and using a firearm, but sometimes the situation may warrant it. Consider keeping a 12-gauge shotgun on hand, and a hand pistol, along with plenty of ammunition. Do make sure you know how to use one; there are plenty of gun ranges you can attend for training.

Martial arts may be something some people laugh at, but self-defense is exactly what the martial arts are all about. One particularly good one to consider learning is Krav Maga, but you could just as easily opt for judo, Taekwondo, or karate – each serves its purpose well.

OPSEC

This comes under self-defense, but it is best discussed separately. OPSEC stands for Operations Security, and it really means to keep things on a need to know basis. In other words, don't go blabbing your plans to all and sundry. Don't let everyone know what stores you have in place because, when the SHTF and people run out of supplies, they'll be at your door, looking to share what you spent precious time and money on.

However, there is another school of thought that you don't have to keep things entirely secret. There is no harm in getting all your neighbors together, explaining what a prepper is, and why they should get in on the act right now. More people need to be educated on what the prepper lifestyle is all about, but what you don't need to do is tell them where you keep your food, your money, what weapons you have stockpiled, where your bug-out location is, and so on.

The thing is, even if you don't tell anyone, or keep it purely within your family, others will soon notice that something is going on. For example, you head to the grocery store, and they have 5lb. bags of beans and rice on special offer or the flour is "buy one, get one free". These are on your list, so it makes good sense to buy them at the special price, but do you really think that no one will notice that you have 6 lbs. of flour and 20 lbs. each of rice and beans? Do you think nobody will notice when the UPS van starts delivering numerous boxes to your door?

Keep things as secret as you need to, but keep it real too.

Preppers Toolkit Checklist

You will not necessarily require everything on this list – much of it will depend on the situation at hand. Use your head, keep things sensible, and remember: You may have to travel on foot to get away from your area; do you really want to be carrying half a ton of stuff for miles?

Water

- A minimum of one gallon per person per day, preferably more
- Portable water filters
- Purification tablets
- Standard chlorine bleach (unscented), 8.25% sodium hypochlorite
- A way of boiling water – gas stove and gas canisters, matches

Food

These are absolute minimum amounts for two weeks:

- 20 lbs. of rice
- 20 lbs. of beans
- 20 cans of vegetables
- 20 cans of fruit

- 20 cans of meat
- Two bags of flour
- One bag of sugar
- One bag of salt
- One pound of oats
- One gallon of olive or coconut oil

Other foods:

- Freeze-dried foods
- Dehydrated foods
- Coffee
- Tea
- Powdered milk
- Powdered eggs
- Canned cheese
- Canned butter
- Instant potatoes
- Pasta
- Canned or packet soups

First Aid

- First aid case/bag – waterproof, lightweight
- Survival medicine/first aid handbook
- Tweezers
- Infrared forehead thermometers
- Antibacterial Q-tips
- Large trauma shears
- Surgical grade toenail clippers
- Scalpel plus spare blades
- Stethoscope
- Blisters plasters
- Band-Aids – various sizes

- Sterile gloves
- Sterile gauze pads – various sizes
- Alcohol wipes
- Bandages – various sizes
- Triangular bandages
- CPR pocket mask
- Steri-Strips (or equivalent)
- Moldable foam split
- Iodine
- Tourniquet
- Sunblock
- Bug repellent
- Burn cream/gel
- Medications and ointments
- Antibiotic cream
- Hydrocortisone
- Antifungal cream
- Antibacterial soap
- Ibuprofen
- Paracetamol
- Tylenol
- Sudafed (or equivalent)
- Antihistamine cream and tablets
- Imodium (or equivalent)
- Throat lozenges
- Oral rehydration
- Prescription medications
- Asthma inhalers
- Multivitamins

Sanitation

- Five-gallon bucket
- Heavy-duty trash bags
- Sealable container
- Cat litter
- Disinfectant spray
- Antibacterial soap
- Hand sanitizer
- Wet wipes
- Biodegradable toilet paper

Cooking

- Portable gas/propane stove
- Gas/propane
- Stainless steel pots and pans
- Cooking utensils – knives, forks, spoons, etc.
- Plates (tin/stainless steel/disposable)
- Disposable BBQs
- Can opener (manual)
- Bottle opener
- Trash bags
- Matches
- Wood
- Power
- Flashlights – one per person
- Plenty of spare batteries – standard and/or rechargeable
- Power/battery bank
- Portable generator – solar or gasoline
- Gasoline – if choosing a gas-powered generator
- Wind-up flashlights

Cash

- A minimum of $1,000 in small bills
- Plastic bags/storage containers

Communications

- Two-way or ham radio – one each
- Spare batteries
- Battery or wind-up radio

Mobility

- Tent
- Sleeping bag
- Space blanket
- Water bottle or bladder
- Portable filter
- Food rations
- Gloves
- Jacket
- One change of clothes
- Warm headgear
- Matches
- Flashlight
- Headlamp
- Basic first aid kit
- Map of the area
- Compass
- Mini shovel
- Ax/hatchet
- Paracord
- Multi-tool
- Knife
- Pepper spray

- Charger – solar or battery
- Whistle
- Goggles
- Copies of your important documents
- Passport
- Titles and contracts
- Address book
- Family disaster plan
- At least $500 cash – small bills
- Prescription medication
- Small mirror

Self-Defense

- Shotgun
- hand pistol
- Ammunition
- Pepper spray
- Knife
- Taser
- Stun gun

Conscious Stockpiling: Items You Should NOT Spend Your Money on

For some people, those who have plenty of cash, prepping is easy, but for those with little spare income every month, it is hard. Most people have been in this situation before, and many still are. Everyone knows that, when bad times happen, they have to start cutting back on what they do, and prepping is one of those things – many preppers have had to put a stop to it and start living off their accumulated stores, just to survive. Having that stock of food can be a real godsend, but it won't last forever, and somewhere along the line, you have to start again.

The first thing to get into your mind is that prepping doesn't need to be expensive, and there are plenty of ways that you can prep for free; you just have to change your mindset to one that says you should never waste what can be used later. This isn't always easy, but drawing up a proper plan, does make it easier.

Survival Food

One thing you can do is opt to use every last scrap of food, saving you money on your food budget. The trouble is that most people don't want to eat the same thing day after day, but you don't have to. Food leftovers are free food for your survival stores, and you can make this work in several ways:

• Dehydrate leftover beans, pasta, and rice

Most people make too much, and most throw away what they don't eat. Stop. Dehydrate them, and when you really need a meal, all you need to do is rehydrate them for a few minutes in boiling water. Once dehydrated, vacuum pack the food for easier storage.

• Canned meat and meat dishes

Did you cook too much chili or Bolognese? Pressure can it following the correct canning process. If you don't have enough for that, store it in your freezer until you do and then can the whole lot.

• Dehydrate bread

If you have leftover bread, rather than throw it out, dehydrate it and turn it into an instant stuffing mix (check the internet; there are plenty of different recipes) or even into breadcrumbs for breading fish, meat, etc.

• Get free food

It can be done. All you need to do is save and stack your coupons. You'll get plenty of freebies and, in some cases, they may even pay you to take them home. Don't go overboard with this; just pick the stuff that you can store long term. And there are couponing classes you can attend to learn how to do it.

• Free fruit

Most yards have fruit trees, and for the most part, much of it is left to rot. Don't be afraid to ask if you can have some. Offer to pay for it, but most of the time, people will tell you to take as much as you want. What then? Make it into jams and jellies, can it, or even dehydrate it.

• Free food samples

A large number of food storage organizations will happily send out free samples of their products in the hopes that you will order more. Even if you don't buy any more, the free samples will go nicely in your bug-out bag or emergency food kit.

• Keep your condiment packs

When you order takeout or head to a fast-food restaurant, grab some condiment packs. These fit nicely into your food kits, but be aware; they don't last all that long, so plan on replacing them yearly.

• Save drink bottles for water

Don't waste money buying bottled water. Simply keep all your drink bottles, wash them thoroughly, and fill them from the tap. The only thing this doesn't apply to is milk containers as milk residue taints the water.

Household and Garden Survival

Couponing can pay off here too; save coupons for toiletries, toilet paper, toothbrushes, toothpaste, razors, and other household items – you can get an awful lot of these for free. And when you visit the dentist, ask for free samples.

• Save your seeds

Be it free seeds in magazines or seeds from plants you grow, vacuum pack them, and save them in a cool, dark cupboard. That way, you can grow your own vegetables when the need arises.

• Grown new plants from vegetable roots

When you buy vegetables or even grow them in your garden, you can regrow them from the roots. Celery, lettuce, onions, even pineapples, and so many other vegetables and fruits may be regrown in this way. Chop off the root end and suspend it in water until you see the roots growing and/or shoots from the top. Then simply plant it in soil and watch it grow. If you stagger your growing, you can have an almost never-ending cycle of free food.

- **Free plants**

Check with your neighbors to see if you can have cuttings or roots from their herbs, and other edibles. They may even give you entire plants if you are lucky. And then just plant them and enjoy the fruits of your labor. If you have an excess of some plants, you can share those or even swap with neighbors.

- **Free mulch and compost**

Many tree trimmers or your county will often have free mulch and compost, but be careful with it – it may be full of insects and pests, not something you want to introduce to your garden!

- **Collect pine cones**

They make great fire starters and burn very hot and quickly. Start collecting them in fall when they are on the ground and store them somewhere cool and dry.

- **Free sandbags**

If there is any chance of flooding, your county may hand out free sandbags. You do need to be quick to get them because they tend to go fast, but they make a great addition to your prepping, especially for the hurricane season.

Shelving and Furniture

Keep your eye on Craigslist, freecycle lists, or for items that people leave outside, free to a good home. You might be surprised at what you can find for free.

Preppers Supplies

- **Free buckets**

You can get these from commercial kitchens; some charge a tiny amount, but on the whole, if you check by their dumpsters, you'll find them for free. Do ask them, but usually, they will tell you to take them away. You can also ask at bakeries, fast-food restaurants, and delis too.

• Popcorn tins

These aren't in such wide supply now, but if you can get hold of the three or five-gallon size tins, they make great storage containers. They are rat/mouse-proof, and you can store quite a bit of packaged food in them.

• Free batteries, tarps, flashlights, etc.

Some stores offer these for free with no purchase needed, but more often than not, you will need to buy something else. It's still a free item that you would otherwise have to pay for, though, so you are saving money.

• Free Samples

Some of these have already been mentioned, but do keep your eye out for internet ads that offer freebies, such as small flashlights, multi-purpose tools, paracord bracelets, etc., especially on survival websites. And provide a company with an honest review of a product they sell, and you may get a coupon for a freebie as a way of saying thanks – this tends to work better with the large manufacturers.

• Free information and Books

Check out the kindle app store for free books. They won't be as detailed as this guide, but you can pick up short survival cookbooks, first aid books, and so on.

These may all seem like small things, but the value soon adds up. There is no requirement to have money in the bank to begin prepping, and you can get so much for free if you know where to look and stop wasting so much at home. Your garden is a great source of free food, too; it doesn't take much to get a few vegetables growing.

31 Essential Prepping Skills to Know

Every prepper and survivalist will have their own priority list of skills to learn, alongside their tasks and gear. Being a prepper really isn't easy, but learning the skills won't just stand you in good stead when it comes to a survival situation; they can also help you after the disaster.

No prepper learns just one skill; you really do need to be a jack of all trades. You must be as prepared as you possibly can be for whatever emergency arises. And face it – no one knows when it might happen, so now is as good a time as any to learn your skills.

So many people think prepping is all about stockpiling food and gear, and while that is a big part of it, it's also about what you know and what you can do.

Beneficial Prepper Skills

There are many skills that you should know, but while every member of your prepper family should have some knowledge about all skills, it is OK to delegate so that each of you learns some useful skills.

Bushcraft

Bushcraft is an ancient art that uses natural resources to survive in the wilderness. It isn't one skill; it's a group of skills that encompass:

1. Foraging for food

This includes knowing what plants you and can't eat, how to cook over a campfire, and how to harvest efficiently without destroying a resource completely. Many plants will regrow if you take a little care, rather than hauling them up by the roots - that's a sustainable source of food. You need to understand what mushrooms you can and can't eat and how to cook them.

2. Hunting/trapping/fishing

Learn how to track and stalk animals for food, how to build a snare and where the best places are to set them. Learn how to hide your scent - remember: Animals have a much better hearing and sense of smell than humans do; they'll know you are there long before you see them. You need to be able to ties knots, make cordage, and importantly, learn how to prepare and cook what you catch. You must also learn how to build and use weapons, such as slingshots and a bow and arrow, all of which can be made from natural resources. When times get desperate, everyone with a weapon will be after the same food sources, and those who prep properly will know exactly what's in their area and know how to catch and preserve it. Another useful skill in this area is fishing - most places have a body of water where some fish are available, and knowing how to catch them and preserve your catch is a vital survival skill.

3. Finding and gathering water

Water sources are more prevalent than you realize; you just need to learn where to look for them. Once you've found your water source, learn how to filter and purify it, so it is drinkable, and if you don't have one, learn how to make a container to collect and carry water in.

4. Building a shelter

This is important; you won't survive long out in the elements, no matter how well prepared you are. You must learn how to fell a tree for materials, baton branches, and find other materials you need to construct your shelter. For example, bark and grass can be thatched or woven to make a roof. You must also learn what materials you can use to both insulate and waterproof your new home.

5. Building a fire

Another very important factor for survival is knowing how to build a fire. Learn the best woods for quick burning and longer burning times, what constitutes tinder, how to build a fire-starting device, such as a fire plow or a bow drill, and how to build a firepit. You should also learn to make charcoal.

6. Navigation

Learning to use a compass or even a watch for navigation is a vital skill, but you should also learn to use other methods, such as the sun, stars, and even landmarks.

There are plenty of courses you can take to learn all this, along with videos and other resources.

Backpacking and Camping

Two very important skills:

7. Backpacking

Not only is backpacking fun, but it's also a great way of getting in shape and learning how to traverse different terrains carrying your survival gear. You'll learn how to carry your bug-out bag properly – the weight should be on your hips, not on your shoulders. You'll learn how to keep the ticks, mosquitoes, and other insects away, avoid injuries to your legs and feet, and how to work together as a group.

8. Camping

Camping is a no-brainer really, and it's much easier to learn than backpacking because you don't need to hike out anywhere. You can learn how to camp in your backyard, and you should already have your camping equipment ready – get out there and start practicing. Stocking up on freeze-dried and dehydrated foods for your camping trips means your prepping skills are not going to waste – you get to find out which ones you like and don't like before you fork out for bulk amounts.

Food Survival

There are several factors here:

9. Baking bread

You should learn how to bake a basic bread over a campfire by hand – no bread maker here! You'll also learn to make biscuits, flatbread, tortillas, and other nice bread. Invest in a grain mill if you are bugging in, and learn to mill your own flour.

10. Bee-keeping

If you are bugging in, learn how to keep bees and extract the honey for food and the wax for making candles. You should also learn how to brew your own beer, not just for your own consumption but for bartering purposes too.

11. Learn to make butter and cheese

If you have livestock, such as a cow, learn how to churn your own butter. It's hard work, but incredibly satisfying when you get to taste your own butter. And you can also make cheese from the milk of cows, sheep, and goats – but remember to wax it to preserve it for longer.

12. Preserving food

You must learn how to can your own food too. Fruits, vegetables, even meat may be canned, and it's even better if you grow and raise your own food. Not only do they make for good meals, but you can also use canned foods for bartering.

Charcuterie is a great skill to learn, the art of curing, smoking, drying, and salting meat and fish. You should also learn how to properly dehydrate food, especially if you don't have the luxury of a dehydrator at your disposal. This is one of the best ways of preserving food, but be aware that it is labor-intensive. Learning to freeze-dry is a better method.

13. Entomology

Grimace, if you want, but learning how to tell the difference between poisonous and edible insects is a great survival skill to learn. For example, crickets are full of fats, protein, and vitamins, vital for energy and health while in the wilderness. As a rule, the bright-colored bugs tend to be poisonous and should be avoided.

Homesteading Skills

Homesteading is all about self-sufficiency and encompasses many different skills:

14. Keep chickens

It's dead fashionable these days to have a few chickens in your yard, and it's a great way of prepping. First, you get a good supply of eggs that you can use for cooking – you can even freeze-dry them for future use. Chickens are natural gardeners too, keeping your land free of insects and producing fertilizer, and, as a last resort, they are a good source of food.

15. Learn to compost

There is a right and wrong way to do it; if you don't know how to do it and what you can and can't compost, you are at risk of illness. And it's important to know how to compost human waste too.

16. Gardening and saving seeds

Gardening is a great skill to learn; not only is it physical exercise, but it also gets you out in the fresh air, and you can grow your own vegetables and fruits. You can grow as little or as much as you want – the more you grow, the more you have to preserve for the future, and gardening is quite simple once you get the hang of it, and it's not expensive. You should also learn to harvest seeds from your plants;

that way, you can keep your crops going for as long as you need. And some seeds are a great source of food, such as pumpkin and sunflower seeds.

Another great tip is to build yourself a walipini. This is an underground greenhouse, where, built right, you can grow vegetables and fruits all year around. Make it big enough, and you also have your very own storage room, out of the elements and out of sight. There are also cases of people enlarging them so they can live in them when the need arises.

17. Learn to knit, sew and crochet

Socks won't darn themselves, and you may not have the option of purchasing new clothing. As such, it's important that you keep what you have repaired and is fit to wear. You can also learn to make new clothes, including socks and slippers.

18. Raise livestock

Apart from chickens, there are other animals you can raise for resources. You don't need to get a cow, and not everyone has the room for one. Instead, consider these:

- **Rabbits** – these take a lot of care, and you must keep the buck and doe apart unless you are breeding them. These give you a great source of protein-rich meat and fur for clothing.
- **Goats** – these are more practical and a bit easier to keep than rabbits and provide multiple benefits: highly nutritional meat, a way of keeping your land clear (they can create a fire break around your property), milk (from which you can make cheese, soap, and yogurt), hide, which you can tan into leather, hair for making into mohair, and lastly, dung (you can use this for compost when it's fresh or as fire fuel when it's dried).

Any of these animals can be a lifesaver in terms of food and useful materials.

19. Making soap

Learning to make soap can save you a fortune in the long run, especially if you have goat's milk on hand. You can even make soap from ash!

20. Shoemaking

One of the lost arts, shoemaking is a very important skill to learn and could save you money – as well as preventing you from having bad feet.

First Aid Skills

There are a couple of main points here:

21. Basic first aid

Taking a basic first aid course is a must – it can be the difference between life and death when there is no access to medical help. You can take things further and go on to learn emergency medical technician skills.

22. Herbal healing

One more thing you can do is learn herbal healing – there are plenty of plants and herbs that can be used as medicine.

Communication skills

One of the most important skills to learn is communication, not just between your prepper family members but others around you too.

23. HAM Radio

Communication is key to knowing about food and water supplies, what condition other areas are in, and in finding out the latest disaster status. Although you need a license to operate one and it isn't the cheapest equipment, it is a vital skill to learn. Plus, although illegal to broadcast, in a disaster situation, you can do it.

24. Morse code

This is an interesting one but a potential lifesaver when all communications are down. With a series of simple dashes and dots, you can communicate just about anything – on paper, visually, via sound, via body language, using flashlights, and much more.

Other Useful Skills

While the above are vital skills, there are other important ones you should learn too:

25. Self-defense

Yes, you could use a gun, but there are other less lethal ways to protect you and your family. You could learn any one of several martial arts, and although they can be lethal, they don't have to be. They do ensure your self-confidence, and they can prove to potential attackers that you mean business.

26. Swimming

Not everyone can swim, but you should learn. The planet is 75 percent water, and there is a good chance that you will need to cross some at some point. You must know how to swim fully dressed, swim underwater, and get yourself out of trouble in the water.

27. Welding

This is a great skill that anyone can learn how to do. It can prove invaluable for repairing vehicles, making weapons, and even creating your own source of energy.

28. Couponing

Learning how to coupon efficiently can result in a ton of savings, and even freebies that can help you in your prepping. Believe it or not, you can even take courses on how to do it.

29. Paracord

Paracord is one of the most popular prepper materials and has a multitude of uses. It's high strength and can be used for lashing together shelters, make-shift furniture, belts, braces, keychain bombs, and many other things. This is one skill you will never regret learning.

30. Whittling

Learning to whittle can help you produce useful items, including a bow and arrow, a slingshot, games to keep the kids occupied, and so on. And you can use the wood shavings to help you light a fire.

31. Bartering

A very important skill to learn is how to barter with very little, trading up as you go. Or using your excess to barter between neighbors, using the money you salted away, and so on. You can barter with just about anything, but one tip is this: Never let on how much you have. People get easily jealous, and that leads to trouble, especially in desperate times.

There are loads of skills you should learn as part of your prepping for when the SHTF and these are the most important and useful.

Top 15 Rookie Prepper Mistakes to Avoid

As mentioned, prepping isn't easy to do, and as a beginner, you are likely to make many mistakes. You can scare yourself to death reading everything you find on the internet about prepping, especially when people talk about things that went wrong. However, proper preparation and attention to detail mitigate most risks – you cannot account for everything, but you can certainly reduce the potential for things to go wrong.

To that end, here are the top 15 mistakes that you should avoid making:

1. Keep Your Mouth Shut! Do Not Tell Others About Your Emergency Shelter

Have you ever watched "Shelter Skelter", an old *The Twilight Zone* episode? Find it on YouTube and watch it - it will tell you why you should keep your mouth shut. In short, a man at a party shoots his mouth off about his emergency shelter. Then, when the emergency sirens go off, every one of his neighbors turns up at his door, smashing the shelter door off. It turns out that it was a false alarm, but if it happened to you and the cause was an atomic blast, you would have received a huge dose of radiation.

The only people you should tell are those that you trust 100 percent and that you want joining you in your shelter should the need arise. The only other people you can talk to it about are other preppers if you have a community near your home. Other than that, keep it zipped and tell no one else.

Desperate people are the most dangerous, and the last thing you need is desperate people trying to get into your last safe place.

2. Not Doing Your Disaster Scenario Homework

Some disasters are common in certain areas, and the biggest mistake you can make is not prepping for what is common in your area. It's fine if you've prepped for all potential scenarios, but if you, for example, prep for the worst case, such as nuclear fallout, and not for tornadoes that are common in your area, then you're going to be out of luck when the next one hits.

Say that your area is prone to severe flooding every ten years or so. You might think that ten years is a long time and the last one only happened two years ago - you've got plenty of time, right? Wrong. With global warming and the crazy weather patterns of late, another flood could happen at any time. And if you're not prepared for it, you could be in serious trouble.

Obviously, you don't want an underground shelter in a flood scenario - you will lose the lot. And what if you live in a hurricane-prone area? You may not have to leave your home, but you should be prepared for water, gas, and electricity supplies to fail, at least for a few days. You also need to be prepared for the temperatures - a disaster can strike at any time, be it the height of summer and or in the depths of a frozen winter.

3. Not Staying in Shape

As a prepper, you need to be ready for anything, and that means staying in shape. You cannot possibly protect yourself, your family, and your property if you can't even make it up one flight of stairs without collapsing at the top.

You live in a fast, busy world, and there isn't always enough time to work out. And then there's the expense of a gym membership to consider. Well, you don't have to join a gym to get in shape. Start working out at home instead. There are plenty of exercises you can do in just five minutes at home. And if you take public transport to work, get off at an earlier stop and walk the rest of the way. Take the stairs, not the escalator or elevator; go for a half-hour walk at lunchtime. So many things you can do, and it all adds up to a fitter, more prepared you. And if your family needs to get in shape too, head out for bike rides and hikes on the weekends.

4. You Don't Have any Training or Survival Skills

It's all very well splashing out on expensive survival gadgets and spending a fortune on prepping – which you don't have to do –, but you also need to learn the necessary skills to use such gadgets to the best effect. Buy whatever you think will work, but take the time to learn how to use it in the event of an SHTF scenario.

Think about it; you get caught out in a crisis, you don't have your bug-out bag with you or any weapons for self-defense. Or you do, and you don't know how to use them. What then? You could rely on yourself to a certain extent, but without survival skills and training, you won't last long in the face of danger.

Provided you are in good shape, learned self-defense, and have a variety of survival skills under your belt, you have a much better chance of surviving pretty much any disaster situation. You don't have to be Bear Grylls, but you do need some skills.

5. Not Choosing the Right Foods for Long-term Storage

You might think any old dried or canned food will do, and there are some strange items that you might see on survival videos that really are not suitable items to stockpile. Take Ramen noodles or Ramen soups – there is more nutrition in a cardboard box! They are not healthy and are full of salt, and unless you have several liters of water spare to wash it down – which you won't have –, you won't last long living on them. Plus, they contain no protein and no other nutritional value.

You must check the labels on the food you store. Too many people pack their kits full of high-sodium snacks – while it is nice to have the odd treat, that's not what this is about. Concentrate on ensuring you store foods with a balance of protein, fat, and carbohydrate – these are the essential macronutrients for health. Having a few salty snacks on hand is fine, but don't go overboard – be sure to do your research before you buy emergency food stores on the internet.

6. You Have Too Many Small Pets

This could be a controversial point. Most people prep for natural disasters and smaller-scale disasters, those that could leave them with no power or water for days or longer, and in these cases, a couple of smaller animals is fine. If you are one of those that believes the apocalypse is just around the corner, though, and you must have an animal, get a bigger one.

Large dogs can help protect you when you are trying to survive, while smaller animals will drag you down.

7. Forgetting to Have Something to Exercise Your Mind

Whether you bug in or bug out, you need to have something to keep your mind ticking over. While food and water are important for survival, don't neglect your mind. Put some books in your stockpile, a mixture of genres, and don't forget to include a few survival books too – these will be of great benefit in an SHTF scenario. Put away books on creating a garden, food preservation, first aid skills, and anything else you can think of that will help in a desperate situation.

You should also store a few board games, packs of cards, puzzle books, and other fun things that will keep you occupied and stop your brain, turning to mush.

8. You Don't Have Fitness Gear in Your Emergency Shelter

This doesn't refer to "the big stuff", like running machines and so on, but you should have some kind of exercise equipment. Not only will it keep your fitness levels up, but research also shows that exercise can stop you from becoming depressed.

All you need is a few dumbbells, a set of resistance bands, and any other small fitness equipment you can think of. That type of gear can even go with you if you need to bug out.

9. You Have Far Too Much Reliance on Electronics

How many prepper videos have you seen where people bury items and then mark them on their GPS? You might think that's a great idea, but what happens if the grid goes down? What happens if a massive electromagnetic pulse (EMP) bursts in your area? That's right, your GPS won't work, and you won't find where you buried your gear.

Do things the old-fashioned way – learn how to read and use a map. People do still use them; not everyone relies on a GPS to get them where they need to go. Invest in a map showing your area and a bit wider out if necessary. You can also invest in a map book, showing the whole country. That way, you can get where you need to go without relying on a GPS.

You should also learn to navigate by the sun and stars – it's not as hard as you think. You can also purchase a small Faraday cage or Faraday bag that will keep your electronics safe in the event of an EMP.

10. You Spend Far Too Much on Survival Gadgets

This is another major mistake by prepper beginners. You start surfing the Net, looking for survival gear, and you get sidetracked by expensive, unnecessary gadgets. How many variations of a knife have you seen? How many different axes or flashlights? At the end of the ay, you should keep it simple. Buy only what you really need and leave the gadgets to someone else. Set yourself a monthly budget and don't go over it.

11. You Don't Monitor Expiry Dates

You should always have water purification tablets on hand and two types of food storage – near future and far future. For the latter, make sure you check expiry dates and rotate your stock regularly. When you buy items, stick an expiry label on it and have things stored in order of their date – this also applies to your water containers, as water will go bad if not stored properly or it gets contaminated in some way. Your water purification tablets will help if your water goes green, so make sure you have plenty of them and that they are always in date.

12. Failing to Stay Organized

This is, without a doubt, one of the worst mistakes you can make. It's one thing to ensure you have adequate food stock, but storing it in a room and then forgetting about it is not good practice – by the time you need it, it may no longer be edible.

You should also plan for the eventuality that you may need to evacuate – not just your home but your town or city. Make sure you have a bug-out plan in place that gives you a fast, safe way out, avoiding huge traffic tailbacks.

13. You Don't Have Enough Water

Most preppers aim for 72 hours' worth of food, but water is a different matter. You might think a couple of cases will do the trick, but it won't. You have to consider at least two liters of water per person per day for drinking, more if the temperatures are high. Then factor in cooking and washing, and you can see it starts to add up. The minimum is a gallon per person per day – bear in mind that you can go for three weeks without food but only three days with no water – make it your top priority.

14. Planning to Bug Out and Not Bug In

Many preppers focus on scenarios where they have to evacuate their homes and travel into the woods to bug out. But not all scenarios require that; when earthquakes or hurricanes strike, the best place you can be is in your basement, safe at home.

At the end of the day, bugging in is always going to be better than bugging out, but few preppers even consider it and don't bother making a plan. Clearly, you need to focus on a bug-out plan because the implications are more serious, but never neglect a plan to stay at home for a few days.

15. Keeping Your Prepping Gear Together in One Place

Another common rookie mistake is storing everything in the same place. It might seem sensible, but really, it isn't. For example, say that you store everything in your garage. When a hurricane hits, your garage is the weakest point – what if you can't get into it to get all your gear out? Spread your gear around your home, so at least some of it is accessible should the need arise. You could even consider a rented storage locker for some stuff – that way, if you are out and can't get back to your home, you at least have access to something you can use.

That brings you to the end of Part One of this SHTF prepping guide; in the next part, you will look at homesteading.

PART TWO: OFF-GRID LIVING

Living Off the Grid: Reasons and Misconceptions

Right now, there are almost two billion people living off-grid in the world, not all of them by their own choice. In the United States alone, more than 200,000 families have made the leap, and that number is growing by the year. Much of that is down to more people becoming conscious of the environment, and some of it is down to preppers preparing for an SHTF scenario – a real possibility, given current events in the world, both human-made and natural.

So, what is off-grid living? The simple way to define it is living in a way that you do not rely on utilities, like water, electricity, sewers, gas, and so on, provided by public services. It means that you generate and provide everything yourself, and there are plenty of ways to do it. You don't have to go totally off-grid right from the start, either. You can start small and build up until you are self-sufficient.

Preppers for SHTF scenarios are doing this right now. Some purchase specific properties, while others are gradually turning their properties into self-sufficient ones. And this is all being done alongside stockpiling food and other survival gear.

Reasons for Going Off-Grid

Now, for the prepper, there are two major scenarios you need to prep for in terms of off-grid living – the hurricane/flood/tornado, etc., that knocks your utilities out for several days, or an end-of-the-world, apocalypse scenario, where everything is gone long term, possibly forever.

For others, the choice to go off-grid is entirely conscious, a decision made for many different reasons.

1. **Self-sufficiency** – that's what living off-grid is all about, not depending on external resources because you can make your own. For the prepper, this is important as it removes your reliance on a system that might collapse at any given moment.

2. **Sustainability** – it's all about living a sustainable life, not draining already stretched resources. You will produce more than consume, not only providing for your family but helping your community too.

3. **Renewable energy** – renewable energy makes all the sense in the world, and it really is common sense against using sources that are not renewable, at least not at the speed that people use them.

4. **It's responsible** – in terms of the environment, at any rate, and everyone has to bear the responsibility to make the world better. The more people who choose to go off-grid, the better people can make the environment, and the more chance there is that their children and grandchildren will have a decent world to live in.

5. **It's practical** – pretty much all, or at least most, of people's resources are reused and recycled, getting the maximum amount of use out of them.

6. **A smaller carbon footprint** – living off-grid means fewer resources are used, and less waste is produced.

7. **Healthy lifestyle** – not only are you more active, but you tend to eat a healthier diet because you grow or raise much of it yourself.

8. **Less stress** – not having to worry so much about earning money to pay never-ending utility bills results in less stress. And because you are happier and more active, your sleep patterns are better, and that also lowers stress.

9. **Going back to your roots** – off-grid living used to be the norm; it's only fairly recently that people have come to rely so heavily on the infrastructure provided for them. Get rid of your dependence on that infrastructure, and you go back to living the way you always did – and it's worth bearing in mind that there were less stress and less illness in those days.

10. **Knowledge preservation** – by going off-grid and living a sustainable life, people preserve knowledge for many generations to come. Living in a high-consumer world means this knowledge disappears, and, eventually, it will disappear forever.

11. **Using fewer resources** – it's simple; when you live off-grid, you create more resources than you use, and you use less of the public resources, benefitting everyone.

12. **No more waste and consumerism** – resource consumption is out of control, along with waste. To reduce it, you need to start creating resources instead of just using them. With a renewable, sustainable lifestyle, you help everyone.

13. **Healthier lifestyle** – plenty of fresh air and healthier food are two great benefits: you spend more time outside, and you grow your own food without using pesticides and chemicals.

14. **Building stuff** – this applies to everyone, but preppers in particular – you get to build and make things that you wouldn't normally think about: buildings to live in, storage places, electricity plants (on a small scale, obviously), water gathering and storage, and so on.

15. **You are independent** – because you no longer have to rely on public services and the system, you become more self-reliant and independent, living in a way that suits you.

Common Misconceptions about Off-Grid Living

For preppers the world over, living off-grid is the ultimate solution for when the SHTF. It makes perfect sense to be independent, to not be in a mess should all services and systems fail in a disaster scenario. However, few preppers really attempt to do it properly, and some try and give up fairly quickly. Much of that is down to some of the most common misconceptions and the fact that people realize it takes a great deal of preparation, and money, to live off-grid successfully. A lack of preparation, coupled with a lack of staying power, means many are doomed to fail – needlessly because it only takes planning and preparation to be successful. Some of the more common misconceptions about off-grid living include:

1. I Only Need a Harbor Freight Solar Kit

Many people think this is all they need to power their house, but can you survive on just 45 watts of power? A laptop takes 30 watts to charge up, and your phone will take five watts, but if you're thinking of running lighting, heating, cooking, and other systems, then one of these won't make the cut.

2. It Will Be Easy to Wash My Clothes

There are so many gadgets for off-grid clothes washing, but none of them are going to make your life any easier. Having help to wash your clothes is always desirable, but it isn't difficult to do a hand wash. All you need is a large sink or tub, and a way of drying them. And that is where true off-gridders differ from those who are just playing – they are more concerned about how to get their clothes dry.

If you have great weather, you can dry your clothes outside, but what if it's pouring with rain? If there's a blizzard? Below freezing? Then it becomes a problem. If you have a big, fancy house, with a big fancy solar heating and electric system, then it won't be an issue, but most off-gridders go more simplistic than that. You will need to invest in drying racks and be prepared for things to take a couple of days to dry off. Alternatively, if you have the room, set aside a room or a small shed outside, set it up with a washing line and wood-burning stove, and dry your clothes that way. Alternatively, if things really are desperate, you can always pay someone to do it for you!

3. Solar Panels on the Roof Are a Great Idea!

While many people think this way, the roof is actually the worst place to put your solar panels. Roofs are hot – heat rises – and that kind of heat actually decreases how efficient your solar panels are. Plus, if you need to maintain them, brush off some snow, or clean them, the roof is high up and awkward to get to.

Stick to ground-mounting your solar panels – you can have more than you can on your roof and benefit from more free power and a better way of maintaining them.

4. You Don't Have any Backups – For Anything

When you generate your own power, you don't have a power company to help when something goes wrong. Sometimes, that's not a bad thing; however, you still need a power source, and that means having a backup – for everything. You should have a backup plan for every source of power you need – cooking, heating, showers, food storage, lights, and so on.

5. Thinking You Don't Need Propane

Propane is an off-grid necessity, especially for those who don't have thousands of dollars to spend on a state-of-the-art solar system. Most standard solar systems are nowhere near enough to power everything you need. You could probably get away with not using it if your heating was wood and you had a heat changer fitted to it, but in the dead of winter, do you really want to wait a couple of hours for the heating to kick in and the water to heat? Probably not – you need propane.

6. Thinking That a Wood Stove is "Living the Dream"

Many off-gridders use wood stoves but think of the work – chopping wood every day to make sure you have enough. And you've got to be somewhere where there is enough wood to chop and live off. Then you have to clean out the stove, build it up – every day. Chop, stack, move wood, cleaning the fire building it – every single day. That's what no one tells you; how hard it is. Your day would be something like this:

Wake up in a cold house every winter morning. Fly out of bed to get the fire going again and stack some logs on it. Oh, but you forgot to bring any in last night, so out you go, in the freeing air to get some. By the time you get home from work, the fire has died down – another cold house. Off you go again, bringing in more wood, building a fire up. By the time you've dealt with the snakes that just love to hibernate in woodpiles and cleaned up the mess you made bringing the wood in, you're done for the day, and you haven't even thought about dinner, a shower, and settling down in front of the television.

Does that sound like living the dream? Or would you rather press a button on your solar or propane system and have a warm house in minutes?

7. Thinking that Solar Tracking is Vital

Many solar newbies make this mistake, fitting pole-mounted systems to let their solar panels track the sun. Want to know a secret? Forget that; add another solar panel to your existing system instead. You'll save a ton of money and generate far more power. A solar tracker may improve your gain by around 20 percent – for a 1 kW system, you might now get 1.2 kW. However, they cost a fortune, at least a thousand bucks, and you need huge footings made of concrete – it will cost you around $1,500 in total. Buy an extra panel at around $250, and you can bump your system up to 1.5kw in one day. For less money, you get more power, and there are no moving parts that can potentially break.

8. Thinking That Propane Fridge and DC Appliance are Worth Buying

Yes, DC is way more efficient, and it does use power to invert DC to AC, but there are other things to consider. Many myths come from once true, now false, information, and the biggest problem is relying on what you see on the internet. There are plenty of old, out-of-date websites, and lots of people take what they read as gospel.

These days, inverter technology has come a long way, and panels cost less. It is not efficient to invert DC to AC, but by adding an extra solar panel or two, what you lose is more than made up for. Weigh that up against what special DC appliances cost and the math is pretty simple.

You also have to consider that many electricians don't like working on DC systems, and a lot won't touch them. Plus, the DC appliance market is small, which tends to mean higher prices. The best option is to go for an AC power system and add more solar panels.

The Realities of Living Off-Grid

For many people, the thought of living off-grid brings to mind visions of apocalypse survivors, MRE food rations, and piles of ammo – not to mention long-haired hermits grubbing around for food.

You do need to consider an off-grid lifestyle when you draw up your prepping plans, but the reality is often far from what you see in the movies. Do it right, and you can have lighting at the flick of a button, toilets that flush, a fridge that cools your food, and warm water for a shower.

Think about the long and distant past – your ancestors lived off-grid! Even emperors and kings had to rely on fire for heat and light and used water directly from wells and rivers. They didn't have the luxury of water and power at their fingertips, and you might not either, not in the event of a serious SHTF scenario.

Some choose to reduce their dependence on the grid, to do their bit to save the environment, and some want to live a self-sufficient life. For the prepper, self-sufficiency is key, but you do need to know what challenges you face as well as the rewards you reap.

The Challenges and Rewards

Right now, you might spend time posting pics of your latest holiday on Instagram, spend the evenings watching your favorite Netflix shows or grab a cheeky latte on your way to work. You may still be able to do all those things, but you'll also have to factor in chopping and stacking firewood, tending to your livestock, growing your garden and foraging for food, not to mention building your home. While there are plenty of challenges, there are also many rewards too.

1. Food Is a Massive Responsibility

It should be your top priority, together with water. You must have a reliable source of safe food, and establishing a pesticide-free source of produce and livestock free of antibiotics and raised humanely is a challenge in itself. You have to set it all up, maintain your garden and your animals, and you have to be prepared to butcher your own meat. And when you've done that, and brought your harvest in, you have to store it in a way that it will last, and not spoil or go to waste.

The other side of the hard work is the reward – a safe source of tasty food. Prepping for survival isn't all about eating tinned and packet food; it's about storing your own meat, eggs, poultry, and produce so that you can have decent healthy meals to eat, all cooked on a wood or solar stove. It doesn't get any better than that.

2. You May Have to Build a House

The modern housing people live in now is not designed for off-grid living. Leave the power off for one week, and you will soon see mold and damp, and you'll probably be growing your own penicillin in the fridge. Many off-gridders, particularly preppers, build a separate house designed to be off-grid. And the reason you see so many off-grid houses out in the middle of nowhere is that building codes in the city ae expensive and restrictive.

This is probably the largest DIY project you will ever undertake but look at in terms of the freedom you will have. You can build your home out of any material you choose, even down to using renewable resources, and you design it around off-grid living. You build your house exactly how you want it.

3. What You Don't Do Isn't Getting Done

When you live off-grid, you don't have the luxury of being passive or procrastinating about things. If you don't do it, it won't get done. You may have to spend the spring and summer months gathering, chopping, splitting, and stacking wood. You'll need to be on top of your garden and your livestock, to bring in as much bounty as you can while caring for them.

Nobody is going to empty your composting toilet, your clothes won't hang themselves on the line, and your line won't buzz to tell you that your clothes are dry, either. And that tasty, creamy cheese and butter? If you don't go milk the cow or the goat, you won't get it; it's as simple as that.

Money may be a concern, and some people continue to work while living off-grid. You don't have to if you live in the right community. You can grow more produce and sell it at the markets, barter with your neighbors, learn to knit and sew to make goods for selling, and so on. Everyone has a skill that they can make some money from, but what you won't have to do is clock-watch – punch in on time, attend a staff meeting and all the other irritating parts of on-grid living. Once you become self-sufficient, living a sustainable, renewable life, your financial needs will be small.

4. Weathering the Storm

Everyone knows what it's like for the power to go out for whatever reason. When you live with all the comforts that you get with a public service grid, you really are at its mercy. You have to wait for the water to come back on, or the power to be reconnected – there is nothing you can do.

Go off-grid, and it's all in your hands. It may be hard work, but you will have the luxury of power, water, heating, and everything else you need for when a major storm hits. When the storm warning is given, the city folk head to the stores and panic-buy; you can light your fire, and sit there with a warm cup of cocoa and watch it roll in, warm, safe, and secure.

5. Not Everyone Will Understand

Your off-grid life will be very different from the way many people live, and while some will be curious about your choice, others will be downright derisive. Your conversation may revolve around a rotational crop and livestock plan while others are discussing the latest *Riverdale* episode – you may never have seen it. You may, for a while, feel that you are cut off from reality until you remember that your reality is fresh fruit waiting to be picked, fresh eggs every day, a warm house in the evening that doesn't rely on the power grid supplying you.

Normal life is what you make of it and, challenging though it will be, it will also be the most rewarding way of life – you will seriously wonder why you didn't do it before.

Homesteading 101

The definition of 'Homesteading' is "any dwelling with land and buildings where you make a home." Sounds simple, right? Today, homesteading is used as a term for those who try to be self-sufficient, although it is still your home.

Different Types of Homesteading

Homesteading isn't just a plot of land with a building on it; there are four different primary types of homesteading:

Apartment Homesteading

Think of a traditional homestead and then pare it down to apartment size. Homesteading is all about self-sufficiency, and you can do that in an apartment:

> • **Container Gardening** – if you have a balcony, pop some containers out there and plant some vegetables and fruits. If your balcony is big enough, you could even have a small greenhouse.

- **Small Livestock** – got a big enough balcony? Ask your landlord for permission and keep a couple of hens or rabbits – fresh eggs and meat every year.
- **Preserve Food** – you don't need a ton of space to do this, just a freezer and some canning knowledge. Even if you don't grow your own food, you can buy it when it's on sale and store it for yourself.
- **Grow Herbs** – you can do this is in the kitchen or on the balcony and have fresh herbs all year round.
- **Make Pantry Staples** – buy things like flour and sugar in bulk and make cookie mixes, pancake mixes, and so on. You can make your own stuffing, sauces, butter, cream, etc., with just a little bit of knowledge and getting the ingredients at the right price.

Small and Large Scale Homesteading

These are the typical homesteads, usually found in rural areas and with land to grow crops and raise livestock. Even in a small garden, you can have a greenhouse, grow vegetables, have fruit trees, raise some livestock, and more. With a larger garden, you can do it all on a much larger scale.

On a small homestead, you may not be able to grow enough food for livestock for winter months, so you may have to purchase hay from a local farmer. A larger homestead will give you the land you need to do this yourself, and you can keep larger livestock too, including cattle as well as goats.

Urban Homesteading

Urban homesteaders have smaller gardens, usually in subdivisions, where they grow some produce, keep smaller livestock, like hens and ducks, and if they can get permission, rabbits, and goats too. All it takes is permission to do it, a creative mind, and the get-up-and-go to make it work.

Basic Homestead Steps

To start a small homestead today, there are some important things to keep in mind:

1. **Plan ahead** – don't just get up one morning and decide to do it. You need a plan and short-term and long-term goals. Are you going to be entirely or partly self-sufficient for food? Are you going off-grid? And more. Try and do it without a plan, and you will run into trouble.

2. **Energy Sources** – if you plan to go off-grid, pick a renewable energy source like solar, hydropower, wood, wind, and so on.

3. **Learn to Winterize** – your home must be livable and comfortable in the winter, so learn how to winterize: cleaning gutters, cleaning woodstoves and pipes, cutting and stocking wood, caging your trees and plants to keep them safe, what to do with your livestock, and so on.

4. **Start Gardening** – if you are going to be self-sufficient, you need to know how to garden. You can, if you do it right, keep your produce supply going all year round. And if you grow too much, you can always sell or trade it. Learn what vegetables and fruits ripen when so you can have a year-round garden, grow herbs for cooking and medicine, and learn about crop rotation and companion planting to get the best out of your garden.

5. **Have the Right Pet** – many homesteaders have a large dog, not just as a companion but for protection for you and your livestock. You can also consider cats to keep rodents and snakes down and minimize the damage done to your garden.

6. **Choose your Livestock Wisely** – chickens are dead simple to keep, and you get eggs and meat from them. Rabbits are a good choice too; bred right, you get a good source of meat and fur. Geese and ducks make great homesteading pets,

and if you have the time and space, a cow or two, goats, sheep, even a pig. All of these create food and other byproducts that will help you in your sustainable lifestyle.

7. **Don't Neglect Tools and Weapons** – a knife is a must as it can be used for multiple jobs. Be aware that you will also need to do work around your homestead, so you will need a variety of tools: screwdrivers, saws, hammers, nails, screws, and so on. And don't forget weapons – a gun for hunting and security is a must, along with plenty of ammunition.

8. **Make Your Own** – learn how to make basic items such as clothes, soap, candles, and so on. That way, you are not reliant on anyone else for your household goods.

9. **Never Waste Anything** – people tend to waste food and water because it's easy to restock, but when you become self-sufficient, you can't do that. Every egg, scrap of meat, vegetable, fur, and so on must be used. Gather all the water you can, use every inch of your garden, and never let anything that can be used for something go to waste.

These are the basics. To be fair, one of the easiest ways to start being self-sufficient is to start with a small garden. You don't have to grow everything from scratch – get cuttings from others and grow things that reseed and shoot up extras every year, multiplying your crops easily. Here are some examples:

- **Raspberries** – these fruit bushes send new shoots up yearly. With just one plant, you could soon have a whole row, although you will need to cull the older ones within a few years.

- **Strawberries** – these send out runners which root themselves, growing new plants. Do be aware, though, that the more runners your plants send out, the less likely you are to have much fruit, so do cut some out, or repot them and sell/swap them.

- **Willow trees** – if these grow in your area, simply snip off a baby branch, put in water, and wait for the roots to sprout. It won't take long to get a few trees going like that.

- **Quaking aspens** – these are constantly dropping babies and multiplying all over the place.

- **Potatoes** – if you bought a sack of spuds and didn't eat them all, leave them to grow shoots and then pop them in some dirt. You can even grow these in tubs indoors, so long as they get natural light and don't freeze.

- **Herbs** – most herbs are simple to grow and require nothing more than digging a bit of the plant up and popping it in another patch of dirt.

Solar Energy and Other Power Options

Going off-grid means choosing the right off-grid power sources. Most people wouldn't have the first clue where to start, and many only really consider using solar power. You don't have to leave your city or hometown to go off-grid, or partially off-grid; with the latest leaps and bounds made in technology, there are plenty of renewable options you can fit to any property.

The following are five realistic renewable sources for off-grid power – the first being the most obvious:

Solar Power

Using solar electricity, you can plug in and power your home quite easily. Most opt for a sun-powered system consisting of photovoltaic panels, batteries, and an inverter. Set up right, these can provide a lot of power, especially if you live in a sunny area. There are no moving parts, and they last for quite a while without needing repair. The downside is the cost. Very rarely is it cost-effective to run your entire home off solar power, and you do have to bear in mind solar exposure – some areas don't get as much as others, and for most people, solar is part of the system, not the whole system.

Most people don't put solar panels on their roof for two reasons – it isn't cheap, and it doesn't look nice. There is also the fact that they need cleaning now and then, and going up on your roof isn't the safest. Some people opt for *solar roof tiles*; they are smaller, look like standard tiles, and are tough. If you are building a new off-grid home, consider putting these on all or part of your roof, especially if it is a single-story home. Do keep in mind that they are not cheap, and an area of 2,500 square feet would cost between $20,000 and $50,000 – you may be able to get tax incentives in some areas.

If you look at it over the long-term, going full solar can see your electricity bill come down to zero or, if you go partial-solar, you can cut the bill by 40-60%, And the tiles do last for 30+ years so, given time, it will pay for itself.

Residential Wind Turbine

The wind is definitely a renewable source, not to mention sustainable, and using a residential wind turbine, you can tap into it for your off-grid power system. These are much smaller now than they used to be and can easily be installed in a residential area.

If you have an acre or more of land and you live in a windy area, it is a great option to consider. A typical system, producing 10kW, will cost from $50,000 to $60,000. It is expensive, but you can save 90-100 percent of your monthly bill, and it pays for itself in six years. Some countries/states even offer tax incentives of up to 30 percent, so be sure to check out what's on offer in your region. Do be aware, though, that if you don't get much breeze in your area, the wind turbine will not move, and that means no electricity. Plus, there are moving parts that wear out and need maintenance, not to mention which also have the potential to fail.

Geothermal Heat Pump

Geothermal energy is one of the cleanest and most sustainable forms of heat energy coming from under the earth's surface and supplying energy all day, every day. Geothermal plants are used for harnessing the energy for industrial use, but now you can use a geothermal heat pump to do the same at home. A geothermal pump is both a heating and cooling system, using the ground to source energy in the cold winters and the using the earth in the summer as a heat sink. You can build it as a separate system or have it integrated with your HVAC; it's your choice,

Geothermal pumps work like a refrigerator – they transfer heat from the earth around your house, using pipes filled with water or antifreeze. These pipes attach to the pump, which becomes a heater or a cooling system, depending on the external temperatures.

Micro-Hydro Electricity

If your property has a running water source, such as a stream or brook, you can consider using a micro-hydro system to produce power. Hydro systems use water, using water flowing from a high to a low place, to generate power, and the micro-hydro systems convert running water flows into rotational energy. That is turned into electricity using a waterwheel, pump, or turbine.

These systems are much easier to build and a good deal cheaper than wind or solar power, but the primary downside is that it can only work in specific conditions. If you don't have that running water source, you can't use the system. If you are lucky enough to be able to use it, you can generate up to 100 times the power that a solar or wind-powered system does for the same start-up capital, giving you an unlimited source of energy. It is more consistent, and fewer batteries are needed for storage because the source is harvesting energy all the time.

Hybrid Solar/Wind System

To go completely off-grid, you could use a system that makes the best of fluctuating weather systems. That is a hybrid system and is far more reliable because you do not depend on a single source for your energy. It also works out cheaper because each source's components are smaller than if you used just one system.

As a rough guide, you could have a hybrid system that generates 7-8kWh per day for around $35,000, double the wattage, and the cost goes up to about $60,000.

With technology in wind, solar, water, and geothermal energy advancing all the time, it is now much easier to install an off-grid system just about anywhere. Systems are smaller, and while the initial layout is expensive, you can expect prices to come down as time passes.

Water Sources, Solutions, and Systems

While generating power is an important part of the off-grid or homestead lifestyle, water systems should be an equal priority. People take water for granted; the average person will not give a thought to using up to 100 gallons per day, much of that wasted. Start living a sustainable lifestyle, and everything you learned will go out of the window.

Having a good, reliable, and safe water source is not a backup plan; it is primary and necessary. Most people don't know where to start, though, and what they need to consider. While it is great if you have a pond or a stream on your property, you shouldn't rely on it. The best way is to set up a diverse system to ensure that you are never left without this precious lifeblood.

This guide won't tell you how to build a system; it will give you an idea of what you can use and the pros and cons of each one.

Off-Grid Water Sources

There are three primary forms of off-grid water: below ground, above ground, and precipitation. If you can, you want access to all of these sources as it will give you the best chance of ensuring you never run out.

Well Water

A well can bring up water from deep in the ground, up to 300 feet or deeper. If you already have a well, fantastic, you're off to a good start. If not, consider having one. Instead of using an electric pump to draw the water, convert it to wind or solar power.

Pros

- A reliable freshwater source
- Sink it deep enough, and it won't freeze in the winter
- Relatively clean, although you should filter it before you drink it

Cons

- It takes energy to get the water, be it manual pumping or another method of powering a pump.
- It's a gamble – you may not hit water in your area
- It is expensive, several thousand dollars
- Your water may have been polluted, especially if you are in a fracking area

Rain Collection

Rain is free water, and you should collect as much of it as you can. Make sure you have enough water storage tubs to collect and store the rain to get you through dry periods. Generally, rainwater is clean, soft, and has none of the chemicals in public service water and the excessive minerals found in underground water.

Pros

- The system is easy to implement

- It's clean and free water, so long as you have an impervious surface for it to land on. Be aware that asphalt roofing can leach chemicals into the water

Cons

- Some areas have restrictions in rainwater collection

- You must have an impervious surface to collect the water, and you will need water storage facilities beside the collection area

Ponds

Ponds are for more than ducks to swim in; they're also a backup source of water. You shouldn't use one as a primary drinking source, but it's a great resource that can potentially save your life and benefit anything that lives on your land.

Pros

- If you already have one, they are easy to maintain

- It fills when it rains and provides a fabulous habitat for plants and animals

Cons

- Pond water must be thoroughly cleaned before consumption

- A pump is needed to transport the water out of the pond and into storage

- Getting it from the pond to your house is not easy

Springs

Springs are a wonderful source of clean cold water, and if you have one on your property, you are richer than many homesteaders and off-gridders.

Pros

- The water is sparkling clean, free, and once you have an infrastructure built around it, requires little care

Cons

- Building the initial infrastructure to harness the water needs work and it may not be in a convenient area
- They can be contaminated by your neighbors, deliberately or accidentally

Off-grid Water Utilization Systems

It is one thing having a source of water, quite another to store and use it. There are several storage systems you can consider:

Rain Barrels

You can buy these or make them from food storage barrels.

Pros

- Install them wherever you have a downspout – when it rains, instant water storage
- Small enough that they can be used on most types of property
- Raised on a platform, you can get a decent amount of pressure from them

Cons

- If you buy them, you are limited to 55 gallons, and you'll need many of them. You can make a chain of them, joining them with overflow pipes. When it rains, and the first barrel fills, it overflows to the second, and so on – the more you have, the more water you collect

Cisterns

Cisterns have been used for millions of years, and you can locate them above the ground or below it. They can be made out of just about any material, even stone, metal, or Ferrocement. Locate them at a higher elevation than the tap, and you have a passive pressure system.

Pros

- You can store thousands of gallons of water
- Bury it in the ground, and it won't freeze

Cons

- Constructing them is intensive
- You need to have a decent understanding of your terrain
- You need a lot of space, and if you want pressure, you need elevation

Lugging Buckets

A high proportion of the world's population still uses buckets to carry water, and you can do it too.

Pros

- There's nothing to break down – just have a spare bucket

Cons

- You need to be reasonably fit
- Your buckets must be spotless
- You should use food-grade buckets

Determining When Water Doesn't Need Filtering

You might think that this is silly and that all water needs to be filtered and purified before you can drink it – but it doesn't. Doing it when you don't need to is just a waste of time, not to mention also a waste of precious resources. This is another reason why you should have a diverse system because it lets you make the right choices for your requirements at any given time. All you need to determine, as far as water goes, is what is for human consumption – the rest doesn't matter.

Take watering your garden or your livestock, for example – raw water is just fine for them. Trees are one of the best water filters there are, and you can purchase a filtering system that makes use of them to clean the water. Alternatively, just use rain barrels – plain old ran water is perfectly okay for plants and animals.

Unfiltered rainwater or well water can also be used for doing your washing – there is little point in cleaning and filtering the water only to chuck dirty clothes and washing detergent in it. Hanging your clothes on the line in sunlight disinfects them – sunlight is nature's disinfectant – and you may not have time to get the washing in before an unexpected storm hits, giving your clothes an extra wash in rainwater.

For cooking, cleaning, and drinking, the water must be filtered.

The Off-Grid Budget: How Much Will It Cost?

Money really does make the world go 'round, and you are going to need a fair stash to get started on your off-grid lifestyle. First, you need to clear any debt you have, then you need capital, and the answer to how much it costs is different for everyone because it depends on how much you do and how far you are prepared to go.

So, these estimates may not all apply to you; focus on those that do and keep in mind that they are only a rough guide.

Property – $0 to $25,000 (average)

You may be lucky enough to find a plot of land going cheap or even free, but keep one thing in mind – there is a reason why. The plot may not be right, and often, free land comes with conditions attached (often not free).

If you are prepared to spend money to get exactly what you want, then expect to pay upwards of $20,000+ for up to five acres. You may get a better deal, so shop around and do your homework first. If you intend to be completely self-sufficient, including growing food, the states that have the best potential are:

- Arkansas

- California

- Florida

- Hawaiian Islands

- Kentucky

- Missouri

- New jersey

- North Carolina

- Texas

Shelter – $0 to $150,000 (average)

Do your homework, and you may even be able to purchase land with a house already on it. If not, you need to factor in road access and the costs of building a home on the land. If you are clever and handy, timber from your own land could be used or get in touch with the local forestry commission – sometimes, a company will pay you to let them clear your land.

Failing that, a contractor will be needed; the average cost of a stick frame house in the USA is between $120,000 and $150,000. If you want to pay more, a rammed earth house will set you back around $200,000, and if your budget is much lower, consider purchasing a manufactured, single-wide dwelling for about $20,000.

Keep in mind that you do not want to get into any debt over this, so start small and increase over time.

Wind/Solar Power – $1,000 to $37,000 (average)

Providing power requires a minimum of one solar panel or a wind turbine and an inverter. This costs around $1,000, but you will only get enough power for a small chest freezer or a solar fridge. Add more solar panels and turbines – try for a combination to ensure you always get power. To power an entire home, you would need a system costing around $30,000.

If you do not want to depend on the grid, you will need storage batters and these cost up to $300 each. A full back-up system will cost around $7,000, and don't forget that your batteries will need replacing every three years or so. Check with local server farms – you may be able to get used batteries a bit cheaper.

A Well – $5,000 to $20,000 (average)

If you are lucky, your land may already have a freshwater source on it, but if you use lake or stream water, you will need to clean and filter it first. If there is no fresh water, you'll need a well, and that involves drilling a large hole and using a pump to get the water out. You may not have to go too far down to find water, in which case it will be much cheaper than having to go deeper.

The average depth is 50-100 feet; expect to pay up to $100 per foot. Then you need a good pump, costing around $800-$2,000, all the plumbing and the electrical work, and water storage tanks, which cost $500-$1,000 each.

If you want more than one well, you could consider purchasing a hydraulic drilling rig, costing $10,000-$15,000, but if you live in a preppers community or an off-grid area, you can always make a little money back by digging wells for others.

Septic System – $2,500 to $5,000 (average)

Properly maintained, a septic system can last for up to 40 years, and they are not too expensive to construct. The biggest cost consideration is your soil – if it doesn't drain well, your tank will be more expensive. Make sure you talk to any contractor bidding for the job about this.

Composting Toilet – $100 to $5,000 (average)

Composting toilets are a must as they break solid human waste into fertilizer in the same way that kitchen scraps are broken down. They use heat, oxygen, aerobic bacteria, and time, eventually producing "humanure". The good thing is that little to no water is required, but they do require upkeep and maintenance.

Purchase a self-contained unit for up to $2,000 or build your own for as little as $100 per bathroom. Alternatively, spend around $10,000 or more for a centralized system that connects several toilets and one composter.

Greywater System – $500 to $10,000 (average)

Greywater systems are used to collect shower, sink, washing machine, and dishwasher water so you can use it in your garden or to flush toilets. These vary depending on the complexity of the system and whether it is being installed when you build your home or added afterward.

A simple system could be as little as $500, while a more complex system, collecting all greywater, a filtration system, and storage could cost $10,000 or more.

Geothermal Heat Pump – $7,500 to $20,000 (average)

These take heat from about eight feet below the ground and use it for heating in the winter and cooling in the summer. Again, the cost will be based on how large your home is, your insulation, how much space you have to install a heat exchanger loop field, and whether you are adding it after building your home or during the construction.

Gardening – $100 to $2,500 (average)

For a family of four, you will need around 1,000 sq. feet of land per person to produce enough fruit and vegetables. Seeds will cost you around $100 per year, but once you get started, you can harvest seeds from some plants.

You will also require a fence, which will set you back up to $1,000 for a woven wire fence supported with chicken wire – chain link costs more. You could also use wood from your land to construct your own fence. You will also require some kind of irrigation system – keep your gray water system in mind when you do this.

If you want to grow nut and fruit trees, grapes, and berries, you will need a variety, and this can cost $15-$100+ per bush or tree. Don't forget to start a compost bin or pile somewhere to help feed your new plants.

Livestock – $1,000 to $4,000 (average) + average monthly cost $300+

Chickens are one of the cheapest to keep, and you can usually pick them up for no more than $10 each. You want to average one laying hen for every person. Then you need a chicken coop – if you have scrap lumber, use that, but do make sure it is secure, or you can buy one for $150+.

A breeding pig will cost up to $3,000, and pigs for meat cost up to $100. Two pigs will feed four people for a year. You also need a pigpen with strong fencing and shelter, which could cost upwards of $500. Expect to spend around $50 per month on food, but you can cut that by adding table and garden scraps to their menu.

A cow will set you back between $1,000 and $3,000, depending on the breed. One cow is enough to feed four people for a year, and cattle food is around $200 per month – again, supplement with garden scraps or have a pasture. You need an acre of land per cow, and fencing will cost upwards of $2,500, depending on the pasture size.

If you purchase a milking cow, you can save on milk, cream, cheese, butter, ice cream, etc., and you can opt to breed, raising calves for money or meat.

Outbuildings – $2,000 to $30,000 (average)

A greenhouse cost will differ depending on size. If you just want a small one for starting seeds early, you could buy a kit for around $750, but if you want to go large and grow vegetables all year round, expect to pay around $10,000. Alternatively, build your own out of materials you can salvage cheap or free.

A barn will depend on the features you want – feed storage, tack room, stalls, height, flooring, electricity, water, and so on. You can build a small one for anywhere between $10-$15 per sq. foot or you can go for a steel frame construction, costing $8-$10 per sq. foot. Expect to spend between $10,000-$20,000 (average).

A chicken coop may be built for free from scrap material, or you can buy one from as little as $150 for a basic four-hen coop. If you want something more special, expect to pay up to $1,500.

If you don't have a barn and you want to keep cats, you can build cat houses out of ice chests for $100 or less, or you can purchase a wooden one for $350+.

Lastly, a root cellar is a great way of storing food. A small barrel system costs less than $100, or you can build whatever size you want – the bigger you go, the more expensive it is.

Maintenance Costs

Once you've spent the money to construct your off-grid home, you still need money to maintain it. Aim to spend around $1,000 a month, although if you are a serious prepper, you've already got most of these bases covered:

- **Food** – even growing your own produce and raising livestock will still leave you needing certain items: rice, pasta, baking ingredients, etc.
- **Household** – light bulbs, heating oil, tools, cleaning products, toilet paper, and so on
- **Gasoline** – your equipment and vehicles will need oil and fuel
- **Taxes** – unless you are in an apocalypse situation and the entire system is down, you still need to pay taxes
- **Insurance** – car insurance, health insurance, even property insurance will be needed

- **Health care** – this is a huge consideration: whether you have health insurance or you put money aside for emergencies, it's still going to cost you

In the final part of this preppers guide, you are going to look at SHTF Survival.

PART THREE: SHTF SURVIVAL

10 SHTF Scenarios: What to Expect and What to Do

Many people believe that preppers are only focused on apocalyptic scenarios and do nothing more than hoard food, water, weapons, and survival gadgets. This is a stereotype that only a small percentage of preppers fall into, and it isn't a bad thing. Prepping, to that extent, can be fun to do and keeps you occupied.

However, prepping should really be about all types of SHTF scenarios, not just TEOTWAWKI scenarios – The End Of The World As We Know It –, and stereotyping just makes it all the more difficult to convince others to start prepping.

Prepping is about preparation to survive, and that doesn't just involve apocalyptic scenarios where death is almost a certainty. It means being prepared for anything, including these top ten scenarios, from financial hardship right up to the apocalypse. Before you begin prepping in earnest for the biggest, make sure you have covered every base, including prepping for realistic, more likely to happen scenarios first; then you can build on that for the bigger ones:

One – Financial Hardship

You may think you have a stable job, you may think your finances are safe, but anything can happen, and that includes a temporary or longer-term shut-down of the banking system.

Unexpected hardship is dangerous, and it can come in multiple forms – job loss, divorce, illness or serious injury, several large unexpected bills for repairs. Anything can throw a spanner in the works. Here is what you need to do to prep for a scenario like this, and if it never happens, then you have a financial fallback for when a bigger SHTF scenario occurs:

> • Pay off your debts and have enough money to cover at least six months' salary. You'll need to do this gradually – pay a bit extra off your debts and put a certain amount aside each week or month to save up.
>
> • Make sure you have medical/health insurance that's up to date, not to mention car, house, and any other important insurance.
>
> • Start stocking food and basic supplies. Aim for a minimum of three months' water and food supplies for the whole family and make sure you do the same for household supplies – toilet paper, toiletries, cleaning products, and so on.
>
> • Make friends. A good community is the best way to survive – be it a church, neighbors, and so on. Build strong relationships that can stand the test of time. Help others when they need it, and they will return the favor.

Two – Natural Disasters (Small-Scale)

Think about your particular area – are there any natural phenomena that regularly occur, such as earthquakes, floods, hurricanes, tornadoes, snowstorms, and so on? These are the first SHTF scenarios to plan for – most people watch the news, and they see how these natural disasters are getting worse by the year. To prep for these:

• Do your research. Find out what occurs in and near your area. Run Google searches, get in touch with your nearest weather resource, ask people in the area – if something has happened in the past, it can surely happen again.

• Do more research. Contact FEMA and other US agencies to get ideas on how to prep and what the emergency evacuation plans are for your area.

• Prepare your supplies. Aim for a three-month minimum supply of essentials – if you did your financial hardship prepping, you already have this. Then, depending on how bad things get in your area, double it. Start early; at the first hint of a disaster, your local stores will run out of stock very quickly. And invest in waterproof containers to store everything high off the ground.

• Make sure your building is easy to evacuate and have a family plan in place. In the event of a bad SHTF scenario, evacuation is the likely course of action. Make sure you have gas in your vehicle, spare gas, bug-out bags with the essentials, and copies of important documents in every bug-out bag. Got pets? Even they need their own bug-out bag, so don't forget to factor them into your plans.

When the SHTF

• Make sure you have a way of following the news, so you are up to date on the latest events. And DO follow the official advice given in your area – if you are told to evacuate, do it. And don't waste time; the last thing you want is to be trapped. If you do need to be rescued, make sure you help the emergency teams, not hinder them.

This might all seem like obvious advice, but take this story of a prepper:

He wasn't a newbie; he had been doing it for years and knew exactly what to do. When the SHTF and his town was flooded, he hung back, thinking he could be a hero and save people. He ended up getting in the way of the emergency services, stopped a whole load of people from leaving because he told them he had more than enough supplies, and they all wound up perched on his roof. They needed to be airlifted; luckily, they were all saved, but it could have been so much worse. Do not be that person. Use common sense and don't be a hero.

Three – The Grid Goes Down

This covers a broad spectrum, but in essence, this refers to a major disruption, in one way or another, to normal supply. The most common one is the electricity supply, and it could be out for days, depending on what brought it down. It could even be a shortage of fuel.

In small-scale terms, this is common in many areas; when storms hit, the power invariably goes out, and if you live somewhere like Alaska, even Canada, snowstorms can take power down for weeks on end. Here is how to prep:

> • Have backups for everything – spare fuel and gas, a generator, and a healthy supply of all the essentials. The best-case scenario is to aim to disconnect entirely from the grid and live off renewable energy sources, much like you do with off-grid homesteading. This may not be possible for everyone, so as a minimum, have a generator and fuel.

> • Make sure you can communicate. When the grid goes down, and you can't travel far, being able to communicate with others is key. Keep a basic mobile phone on hand, fully charged – the old type that lasts for days on a single charge. If the mobile system fails, make sure you have a radio setup.

Four – Civil Unrest/Violence/Crime

This is not an uncommon scenario – you only have to look at the news to see just how much civil unrest and rising crime levels there is in the world right now. Could it happen to you? Yes, especially if you live in a highly populated urban area, so here is how to prep for it:

- Make sure you have plenty of food, water, and essentials in stock. During periods of unrest, you don't want to leave your home unless you absolutely have to. Be prepared to hole up and dig in for a while.

- Make sure you have a good neighborhood watch scheme; it's your first and best defense against riots and violence in your area. That doesn't mean not calling the police when trouble starts, but it does mean that you are all watching each other's backs. When things are particularly bad, you could set up a schedule, so your street is being watched at all times. Make sure you have a way of communicating with one another and make plans for what to do if the unrest or violence gets to your street or somebody attempts to enter your home.

- Make sure you have some basic home defense in place, but DO NOT turn vigilante and start firing at anyone who comes near you – that's how innocent lives are lost. Even if you shoot a real criminal, there's a high chance that attention will turn to you in retaliation. If having a gun helps you feel better, then have one, but don't use it unless you absolutely have to. Instead, concentrate on home defense – a tripwire attached to an alarm of some sort often deters small-time looters and criminals. Have a good alarm system and a way to call for assistance if you need it.

- You should also be prepared to barricade yourself inside your home should things get bad. Have shutters over your window, block doors and windows with furniture, and stay away from the windows and doors in case of stray bullets – do leave yourself a means of escape, though – just not a main, obvious entrance.

Five – Economic Collapse

This isn't quite so likely but has happened and can again. This could mean anything from economic depression to extreme hyperinflation. This would lead to the financial hardship talked about earlier, and it wouldn't be easy to get supplies. In that case, your financial planning prep will see you through.

If things are more extreme, like all financial systems collapsing, any money you have in the bank is useless, trade will be impossible (to start with), and you won't have anything – no money and no supplies – in a world where everyone is panicking.

Unless you prep properly, if the economy collapses, you will be most likely die from violence or a lack of basic essentials, so:

- Stock up on food, and plenty of it.
- Make sure you can produce things, like tools and food, that you can trade. If you can grow food, do it – you're going to need it.
- Be prepared to defend yourself, your family, and your home.

Six – Cyberattack

This is becoming more and more likely as each year passes, given the increasing scale of cyberattacks seen now. In a world that is highly connected, the total failure of communications and servers would be a total disaster. What would happen if communication satellites, connected devices, and the internet failed?

Initially, law enforcement, medical care, and trade would break down, accompanied by mass panic. Most governments are prepared for something like this and would likely regain control – eventually. The first weeks or months would be utter chaos, though, and it is down to you to make sure you can survive. Here is how:

- As well as a decent supply of food and other essentials, make sure you have a good stock of information. This refers to copies of your all-important paperwork, certifications, contact information for important contacts (handwritten, not stored on a phone), books that contain useful information you won't have access to online, and so on. You should also keep an up-to-date first aid book and survival books if you can – if not, print it all out from the internet and keep it in waterproof document wallets.

• Have a radio. If all communication satellites collapse, it will be invaluable. Not only is it your way of keeping in contact with others, but it is also the most likely method the government will use to transmit information. Consider amateur ham radio as a hobby; not only is it fun, it could be a lifesaver.

Seven – Terrorism/War

Once, this wasn't a likely scenario, but now it is. Primarily, this refers to persistent terrorism or large-scale wars that could put the whole nation in a state of war. Not nuclear war, not yet anyway, but bombs and guns being used in urban areas, knocking out travel and communication networks.

While it may not be likely in your area, you should be prepared for it anyway. The only real prep you can do here, aside from keeping your supplies up, is listening to the news – do NOT, however, take all news as being a signal that war is going to break out. Be assured: you will know if it is going to happen.

Eight – Biological Weapons/Superbugs

This covers anything from an Ebola outbreak to the use of anthrax and other biological weapons. The first is more likely than the latter, not least because of the sheer level of resources needed for a biological weapon capable of killing a whole nation.

On the whole, modern health care and sanitation have people well covered for things like superbugs, but they can still happen, and you should still prep:

- Make sure you have a supply of clean water. Dirty water is often the cause of disease outbreaks and how they are spread. And, face it – if you were going to target a nation, what better way than infecting the water supply? Make sure you have an independent water supply and a top-notch filtration system.

- If you can, have an airtight room that has a great filtration system to ensure you can breathe properly. If not, go old school and have gas masks on hand – do check the rubber seal has not cracked, though.

- Make sure you have excellent sanitation in the event that the sewage system fails. A septic tank will see you through for a while, but you must have a backup in case things go on for longer. If you have a bug-out shelter, ensure it has a good toilet pit, dug away from your water supply (downwards), and away from your shelter. Also, make sure you have a decent supply of cleaning and sanitizer materials and have a quarantine plan in place should any of your group come into contact with a superbug.

Nine – EMP

An electromagnetic pulse (EMP) is fast becoming a major concern for preppers. An EMP is a huge energy wave that, in theory, could take out every electronic device. People cannot speculate on the likelihood of this happening, and although a solar flare could potentially do it, it isn't likely. However, it is something you should consider, and preparations are the same as for a cyberattack or grid-down scenario.

Ten – Nuclear Blast

While this was once a big fear, especially during the Cold War, it isn't these days. However, with rising tensions among the world superpowers, the threat is always current. You must also consider the number of major nations that use nuclear energy, and with human error a very real possibility, there is always the potential for another Chernobyl.

You further have to consider nuclear missile attacks – Iran, Korea, Russia, to name a few, are more than capable and ready to launch one at the USA, and that makes it a real threat to be prepared for:

> • Make sure you know exactly what to do in the event of a nuclear blast. Your primary priority is shelter and staying away from windows. Obviously, it would be ideal if you had your own nuclear shelter, but most people won't. So the immediate danger is the actual blast – you should have a room prepared with a radio, water, food, blankets, and first aid supplies. It must have no windows, and you should immediately cover your ears and eyes.

> • After the initial blast, you will have several minutes to get to a shelter before the fallout begins to hit the ground. Although they don't offer a huge amount of protection, have some potassium iodide pills and take them immediately.

> • Stay inside, and have the radio on for official announcements.

> • You should have a radiation face mask for each person and a radio in your bug-out or bug-in shelter.

> • Make sure you know where you can shelter at work, home, and everywhere along your commute route. And make sure you have a way of communicating with family and friends.

> • If you can, build a shelter. You can build something basic in your own home. You don't have to go to the expense of a full-on nuclear fallout shelter.

There is one more scenario to consider, although it is extreme. It is TEOTWAWKI; in other words, the collapse of civilization. This means the government has collapsed, along with law and order, not to mention morality, across a nation.

This is unlikely, and if it does happen, it won't be for long. The human race has resilience, not to mention a mostly stable society. Throughout all the disasters recorded through history, civilization has never totally collapsed, at least not long term.

That said, anything is possible, and you should be prepared for it. Here is how:

- You need to be prepared in every way mentioned for every potential SHTF scenario, but you must think longer term than just a few months.

- Make sure you have enough items to barter with and learn all the skills you possibly can to help you. You won't be able to store enough food for a lifetime, but you can learn how to be self-sufficient.

- Learn a few new languages. Once civilization has collapsed, people are likely to travel across the globe looking for somewhere to live. Even with no trains, planes, or ships, most people will find their way anywhere, just like they did many centuries ago.

- Learn to teach. You may not have kids now, but they're going to be the primary way of ensuring the survival of the human race. Learn how to teach kids everything they need for survival – not just survival skills, but languages, history, literature, pretty much everything they need to make civilization work again.

You cannot possibly prepare for every potential scenario, but you can be prepared for almost anything. Start basic by prepping for disasters in your own area and work from there; eventually, you will be prepared for most things.

SHTF Evacuation

If you reside in an urban or city area and are interested in prepping for when the SHTF, you will probably already know that getting out of a city is harder in an emergency than it is to escape the suburbs or a smaller town – you only have to look at events in the news to see how true this is.

For example, in 2015, Belgium was locked down for several days. Tanks patrolled the street, and people were forced to stay in their houses. Why? Because of one terrorist running rampage following the November 2015 attacks in Paris.

On New Year's Eve 2015-16, in Cologne, Germany, a mass sexual assault took place, with more than 100 women filing complaints – all on one night and carried out by a group of immigrants (confirmed).

Riots and terrorist attacks are not the only scenarios that might require you to bug out, but whatever the scenario is, you can guarantee one thing – evacuation will not be easy. A combination of checkpoints, traffic jams, and riots will make it hard to get out, and while those who live in smaller towns or rural areas may have made plans to bug in should an SHTF scenario arise, those in the city may not be able to.

What you can do is have an evacuation plan in place to give yourself and your family the very best chance of getting out safely. The plan you draw up must have several aspects that cover the following:

1. Assess Your Situation

Before you can even begin to make your plan, you must know your current situation. Everyone has different homes and family lives, and that means different liabilities and assets. Some people rent out their homes. Some live in city high-rises. Some have children, live alone, have pets, or elderly parents living with them. You might have an income of $30,000 per year or $130,000.

You must assess your own situation, so you know what you are working with; planning properly now will save you hassle later on and will ensure your evacuation plan is sound. Some of the things you need to ask yourself are –

- How many people are covered by the plan?
- Are there small children, older people, or those who have a handicap of some kind?
- Are there pets in the household?
- Are you in a remote area, mid-city, downtown?
- Do you have somewhere to go?
- Do you have a means of getting there?
- Do you know when you need to leave?

With those questions answered, now there are four more important ones to answer when you draw up your plan –

a. Do you NEED to go?

When you draw up your evacuation plan, you should determine if there is a need to go. Many people opt to stay put, bug in, and defend their home, regardless of the situation. Indeed, many serious preppers will tell you that it is best to bug in for many SHTF scenarios, but there are always times when it's necessary to evacuate. It is down to you to work out when you are safe to bug in, or you need to get going, and that means setting criteria to make your decision by.

b. Do you know WHEN to go?

You may have specific criteria that determine if you need to go or not, such as specific scenarios that require you to bug out. You may opt for bugging out when the grocery stores empty, with no chance of being restocked for some time.

The key point is to make sure those criteria are defined clearly; then, you can determine WHEN to go, should any of those scenarios happen, or are a threat. Leave too early, and you might find you didn't need to leave. Go too late, and you might find it tough to get out of the city; everyone will be trying to get out at the same time, and exit routes will be clogged. You might even find the military has stepped in and shut the roads down.

c. HOW are you getting out?

Your evacuation plan has to be ready at the drop of a hat, and that means ensuring the following:

- Every member of your evacuation group needs a bug-out bag, even pets.

- Everyone should have everyday items with them all the time – you might be surprised at how many survival tools a keychain or wallet can hold: mini flashlight, multi-tool, mini first-aid kit, weapons for self-defense, whistle, and even something to start a fire with.

- You need hardcopy maps of the entire area. Mark every exit route on them clearly and be aware that main routes are likely to be clogged, so make sure you learn the alternative routes. You must all know at least three routes out of the city to your destination, including power lines and railways, especially if you need to get to your bug-out shelter in a hurry.

- Make sure your bug-out location is near, within 20 to 100 miles of your home. If you are relying on a car, you can be the maximum distance out, but if you need to go on foot, you won't want to be far away. You could also consider using bicycles instead of going on foot – it will be hard work, but you will cover more ground. If you can get to a friend in an unaffected city or town, make sure everyone knows where it is and give them a phone number and address – that person may also be an emergency contact.

• Make sure the vehicle you are using to get away has plenty of gas, no less than half a tank. At the very least, you need enough to get to your emergency location.

• Practice your evacuation plan with everyone involved – several times. When the time comes, you don't want any holdups.

Additional Tips

As well as everything discussed, there are a few other things to consider:

• If you can, bug out at night. If the SHTF scenario is a war or a riot, you tend to find that the quietest times are 2 AM to 5 AM. You shouldn't run into any traffic problems, but make sure you know where security checkpoints are so you can avoid them.

• If the situation is urgent enough that you cannot wait, decide immediately whether to use your car or go on foot/bicycle. And be ready to ditch your car at a minute's notice and continue your journey on foot.

• Have supplies and valuables ready to throw in your car when you go.

• Make sure there is a fully loaded bug-out bag in the car – everyone should have their own ready to grab as well.

• Be careful as you leave – depending on the situation, the key is to remain invisible. If you travel at night, try not to use flashlights. If you go by day, do it in a way that won't attract attention.

- You cannot take everyone with you, so don't try, not unless you have infinite supplies and room. It sounds awful, but if you try to be a hero, you could be risking your own life as well as others

- Expect you and your family to become separated at some point. Ensure everyone has a walkie-talkie, a cell phone with extra batteries, a charger, area maps, and knows the set meeting points, as well as the bug-out location and how to get there, regardless of the mode of transport.

WHERE are you going?

Not everyone will have a comfortable bug-out location with plenty of food and water. If you don't, do not attempt to bug out unless it really is the only option, and it's safer to take a chance in the wilderness than stay at home or stick to the roads. Only you can determine the situations that may require this.

If you have no alternative location to go to, consider a national park or a nearby campground instead.

Can Anyone Be 100 Percent Ready?

Not really, but that isn't an excuse to slack off on your preparations. Everything you do will help you prep for an SHTF scenario, even if it's just buying a wind-up radio or doing a first aid course. The important thing is that you start and that you do a little towards your plan whenever you can. And remember – knowledge and planning are far more powerful than any survival gadget you may spot on the internet.

Medical Care During SHTF

If there is one thing many preppers do wrong, it is not having a decent first aid kit. Most opt for a $30 kit from the drug store or department store and think that is sufficient. It isn't. Not when you are thinking of an SHTF survival scenario, in any case.

The only way to ensure that you have what you need is to buy a full trauma kit or make your own. For survival purposes, your first aid kit has to have two things: depth and breadth – and shop-bought kits don't offer either.

What does that mean, depth and breadth? Starting with the latter, breadth means having the right gear to treat many different injuries, from minor to serious trauma injuries; for depth, it means having enough in a kit to deal with several different serious injuries.

You need to keep in mind that medical facilities are likely to become overcrowded and unable to function properly. You may not even be able to get to a hospital or clinic, and having that full trauma kit may well save a life or two.

The most important thing to remember is that first aid is only that, and unless you are a surgeon or a trained paramedic, all you can do with serious injuries is to stop them from getting worse. And some medical supplies, in the wrong untrained hands, can do more harm than good. That is why you should never attempt any kind of assistance that you are not trained for.

Your Kit

The obvious place to start is with your kit bag. It must be something that your kit can be properly organized in, perhaps a case with several divided sections or compartments. A big fishing tackle box is ideal because there are plenty of organizational spaces for small and large items.

Next comes the contents of your kit, and it is recommended to avoid cheap supplies; go for high-quality medical supplies because it can make all the difference when you need to treat someone.

Hygiene

This is one of the most important things to remember – many wounds are not likely to kill anyone unless a major organ has been damaged. Typically, people die from wounds if they bleed out or if the wound has become infected, and while a bandage may stop bleeding, it won't stop an infection. This is why wounds and the area around them must be kept clean, and that means stopping yourself from adding bacteria. Make sure you are clean before you attempt to clean a wound and make sure you are protected from any infections disease the person has.

- **Antibacterial hand sanitizer** – wash your hands before you touch a wound

- **Sterile gloves** – they must be sterile, and you must put them on before you treat anyone

- **Medical face mask** – to stop you breathing germs into a wound

- **CPR mask** – protects you and the person from infection should CPR be required

Flesh Wounds

These are what most of your first aid kit is going to treat, and it could be something minor, such as a cut, to a severed limb. Your goal is to stop it bleeding and provide the wound with protection. The size and severity of the wound will determine what you do.

- **Irrigation Syringe** – to clean the wound before bandaging it. Squirt clean drinking water into the wound
- **Alcohol wipes** – to clean the wound and the surrounding area (you'll need a large supply of these)
- **Antibacterial ointment** – for putting on the wound to prevent bacteria causing infection
- **Steri-Strips/butterfly closures** – used for closing open or separated wounds to protect them, to slow or stop bleeding, and encourage healing
- **Clotting agent** – something like QuickClot or Celox, designed to help clot blood. Some bandages are fused with it, or you can buy crystals
- **Cloth adhesive bandage strips** – flexible, so they don't come off so easily
- **Knuckle bandages** – not the easiest thing to bandage but these work by keeping things clean and moving with the fingers
- **Fingertip bandages** – another tough area to bandage
- **Large bandages** – sterile gauze pads; keep a variety of different sizes for different sized wounds. These do not contain any adhesive
- **Medical tape** – to keep the bandages in place
- **SWAT-Tourniquet** – great as a tourniquet, but when half-tightened, they act as a pressure dressing to reduce bleeding

- **Israeli Bandage** – a combat bandage with a clotting agent and a wrap to hold it in place; it can also be used as a pressure dressing

Broken Bones

These are typically fractures or compound fractures, the latter of which is identified by the bone protruding through the skin. In these cases, both the fracture and the wound must be treated, while with a simple fracture, only the bone must be treated.

- **Splints** – you can use almost anything as a split, but a SAM splint is one of the easiest; constructed of soft aluminum and foam rubber, it can be cut to size and is malleable
- **Elastic bandages** – you might know them as Ace bandages, required to hold the splint in place, treat sprains, and ligament injuries
- **Combat cravat** – another bandage; triangular in shape and used for making slings

Must-Have Medical Tools

As well as medical supplies, you need tools for both simple and serious injuries. Even if you are not trained in using medical tools, these can make all the difference:

- **Medical Scissors** – to cut bandages and clothing off (make sure you get a good pair)
- **Fine-pointed tweezers** – to remove gravel and splinters
- **Jewelers Loupe** – a small magnifying glass that the eye socket holds in place – useful for finding fine splinters, thorns, and so on
- **Hemostats** – should you have to work with a severed limb, these close off veins to stop blood pumping out
- **Tourniquet** – the one-handed type is best, just in case you need it on yourself
- **Stethoscope** – not an absolute must

- **Blood pressure monitor** – when blood pressure drops, it could signal internal bleeding
- **Blood sugar monitor** – self-explanatory: use it on people who are shaky, are not thinking clearly, suddenly lose strength, or feel as if they're going to faint
- **Thermometer** – an increase in temperature indicates fever and potential infection. In-ear thermometers are the best ones or the forehead strip type
- **Eyecup** – for cleaning eyes, used with saline or sterile water

Other Things You Need in Your SHTF Kit

There are so many things you could put in your kit that the big question is: Where do you stop? You can go as far as you want, but these are the must-haves:

- **Instant ice packs** – good for sprains or any non-bleeding injury to reduce swelling; you can also use ice spray
- **Tegaderm** – a film dressing for use over burns and rashes to hold the medical cream in place
- **Benzoin** – used for cleaning areas around wounds to ensure the bandage will stick
- **Lidocaine** – local anesthetic: injection or topical use for reducing pain in wounds
- **Anti-inflammatory medication** – things like ibuprofen or diclotard, anything like that will do the trick

Tips for Treating Wounds

Try to get into the frame of mind where you treat all wounds AS IF they were life-threatening. A mnemonic, ASIF helps you to memorize basic treatment for wounds and lacerations:

A = Amount of Bleeding

Is it a venous or arterial wound? Arterial blood is brighter and streaming or pulsing out. It must be stopped with a tourniquet or compression bandage. If you don't have one, create one by placing a sterile gauze over the wound and apply some pressure – if the person can help, get them to apply the pressure. Elevate the wound if you can and wrap an elastic bandage around it – start from where the wound is closer to the toes or fingers (distal) and work toward the heart (proximal). This will stop blood pooling in the patient's extremities. Do not wrap the bandage tighter to get more pressure – turn it a half-turn, wrapping it over itself, over the wound, and continue to wrap.

S = Shock

Many injuries create shock, even psychogenic shock at times. Reassure the person, keep them warm and calm, and as comfortable as you can. Never say anything meaningless, such as "It's going to be all right." Be supportive and strong; the more you can reassure the person that you are helping them, the more their body will focus on healing without stress, adrenaline, and other factors that can knock off the physiology of balance.

I = Irrigate

If the wound is not arterial or life-threatening, and you can get into it, clean it out. There are plenty of germs just waiting to get into it, so clean as quickly and thoroughly as possible.

F = Functional/Further Damage

If you can clean the wound, inspect it to see if there is any functional damage. Have any major vessels, tendons, or nerves been severed? Can the person move their fingers or toes? If there is any damage, you need to treat that too. If, for example, an extensor tendon has been severed, surgery is required to heal it. You cannot do that. If it was just nicked, you must stop the person from extending the specific digit so it can heal.

Stitches

Many people think they can stitch a wound, but in primitive, remote, or post-disaster environments, stitching a wound is irresponsible. In many simple wounds, suturing is not necessary and can cause infection. Use your bandages to keep the wound clean, but leave it to heal on its own. Use Steri-Strips if you can, and tight bandaging is okay for some wounds. Unless you are medically qualified to do it, steer clear of suturing.

Infection

This is likely to be the biggest issue you will face in post-disaster scenarios. Treating an infection is easier if you can catch it early enough, and that means knowing what signs to look out for. Every wound will show inflammation to a certain degree, and there will almost always be some infection. Some of the key points to look for are:

- **Redness** – inflammation causes a certain amount of redness, but infection creates a whole lot more because more of the tissues are inflamed. It will often be a much brighter shade of red too

- **Swelling** – again, inflammation causes some swelling, but the swelling caused by an infection is typically down to pus, and if touched, causes sharp pain. These usually drain by themselves

- **Pain** – inflammation causes more of an ache, but infection causes sharper pains. This is typical of infection if it hurts while that part of the body is being rested or a much sharper pain is felt when moving after being still for a while. Pain caused by infection could be local to one area, or it may be more spread out than inflammation pain, which tends to be only around the wound area

- **Pus** – otherwise called exudate, pus indicates an infection and not inflammation

- **Fever** – this is one of the more serious indicators that infection is present and has gone further than you want it to be in a post-disaster scenario

- **Streaking** – if you see red streaks following the veins, it signals a serious infection

How to Deal with Infection

Typically, infections are treated with antibiotics, but you may not have any on hand, or you may not have the correct type. What if you give antibiotics and the infection doesn't respond? Or, as is the most common case, you may not have the expertise and training to know whether an antibiotic is working or not, or whether the person has an allergic reaction to it.

No matter which perspective you come from, the first thing is to clean the wound. If infection sets in within the damaged tissues, again, clean it thoroughly. Activated charcoal is one of the best ways, not to mention the most efficient, for cleaning infected wounds, and you can buy it in tablet, capsule, or loose powder form.

Using Charcoal

Using clean drinking water (distilled is best), add it to the charcoal to make a paste – tablets may be crushed and capsules opened and emptied. Charcoal bonds weakly to water, and when used in a wound, it will pick up anything it comes into contact with. Basically, charcoal is a micro-sponge, cleaning and absorbing bacteria, toxins, and dead tissues that feed bacteria.

Once you have mixed it, apply it into the wound and around it. Put a gauze pad over the top and secure it in place in whatever manner works. Within a few hours, you should start to see changes in the state of the tissue. Change the charcoal mixture every few hours until the tissue is no longer carrying an infection.

One thing you must understand is that skills and knowledge are more important than having the biggest survival first aid kit. It does not matter what is in your kit – it is knowing how to use it correctly that counts. Every person in your prepper team should attend a first aid course and keep their training current too. And you should always have an up-to-date first aid booklet in your kit as well.

Additional First Aid Checklist

Combine this list with the previous one given to you:

- Antibacterial hand sanitizer
- Sterile gloves
- Medical face mask
- CPR mask
- Irrigation syringe
- Alcohol wipes
- Antibacterial ointment
- Steri-Strips/butterfly closures
- Clotting agent – QuickClot or Celox, etc.
- Cloth adhesive bandage strips
- Knuckle bandages
- Fingertip bandages
- Large bandages
- Medical tape
- SWAT-Tourniquet
- Israeli Bandage
- Splints
- Elastic bandages

- Combat cravat
- Medical scissors
- Fine-pointed tweezers
- Loupe
- Hemostats
- Tourniquet
- Stethoscope
- Blood pressure monitor
- Blood sugar monitor
- Thermometer
- Eyecup
- Instant ice packs
- Tegaderm
- Benzoin
- Lidocaine
- Anti-inflammatory medication
- Charcoal powder, capsule or tablet

The Bug-out Bag: Surviving SHTF On the Go

The BOB – your bug-out bag – is an essential part of prepping for an SHTF scenario, and it is nothing more than a kit that you can grab and go. It is an all-in-one emergency survival kit, and every member of your prepper family must have one, even your dog. The bug-out bag is designed to be carried and keep you alive for at least three days following an SHTF disaster, but for it to do that, it needs a few essential items.

Do this right, and while everyone else is panicking, running around like headless chickens, you will be way ahead of them and able to sleep, provide for you and your family (with safety, food, water, the means to hunt), wash, and communicate. It all comes down to what you put in your bug-out bags and how prepared you are.

If you are new to prepping, you might be a little overwhelmed by the bug-out bag, especially as you need more than one – one for each member of your group and at least one spare in the car. And if you have a bug-out shelter somewhere, put another one there too – it can be the difference between life and death.

Most preppers continually tweak their bags to make sure they have got everything they need. The hard part is knowing where to start, so here are the seven basic types of gear you need in your bag – once you have these, you can tweak them to your heart's content:

1. Water

Water is a basic survival requirement. You can only go three days without it, and when the SHTF, it soon becomes one of the most valuable commodities.

As a bare minimum, you need one liter per person per day – aiming for a minimum of three days, each bag needs at least three liters of water. You will also need a water purification system to purify other sources of water you come across. It can be a small camp kettle and some iodine tablets or a full-on water filter. Have a collapsible backpacking bucket so you can easily collect more water, and a pack or two of coffee filters to make your filtration system last longer.

2. Food

For a three-day BOB, you can get by on energy bars and freeze-dried meals – these need boiling water. These products don't weigh anything and last for ages. In the long term, your bug-out shelter will need a larger supply of food.

3. Clothing

The clothes in your bag should be no different from what you would need for a weekend hiking/backpacking break:

- Sturdy shoes or boots
- Long pants (not jeans)
- Two pairs of socks (not cotton)
- Two shirts – one long, one short-sleeved
- A warm waterproof jacket
- Long warm underwear
- A hat
- A bandana

This list could go on, and many preppers pack at least twice the amount of clothes – this list should see you through the first three days, though. One thing you must do is pack according to the weather.

4. Shelter

For three days bugging out, you need shelter and somewhere dry and warm to sleep:

- A tent or a tarp you can set up
- A ground tarp and a sleeping pad
- A bedroll or good sleeping bag

If you opt for a tarp instead of a tent, make sure you take whatever you need to set it up.

5. First Aid Kit

As mentioned earlier, what you need as an absolute minimum in your bug-out bag is a first aid kit. It's recommended that you make your kit – don't just buy one off the shelf and think it will be enough. Best case, the contents will be cheap, and worst case, you won't have what you need, but you will have a load of stuff you won't need.

Building your own familiarizes you with the kit and how to use it too, which is essential when you are bugging out.

6. Basic Gear

These are the things you cannot do without but don't fall into any other category. You do not have to follow this list exactly, just use it as an idea of what you should have.

- A minimum of two ways to shelter from the rain – a poncho, coat, tent, etc.
- At least three ways of starting a fire
- Something for chopping firewood
- A small backpack stove, fuel, and something for boiling water for your meals
- A minimum of two good flashlights and plenty of extra batteries

- A survival knife

There are other things, and this guide detailed earlier what you should have in your bug-out bag.

7. Weapons

You may find yourself in a lawless situation, and in times of desperation, people do desperate things. Expect the worst, and make sure you are well prepared for defending yourself and your family. You should have a firearm of some kind, whatever you are comfortable using – and one that you have had some training and practice in using too. Don't turn vigilante, though; only use the gun if it is absolutely necessary.

You can also use your survival knife as a weapon if you need to, and even carrying a club or a large walking stick can be a deterrent. The long and the short of it is: Have several defense options and be prepared to use them in a desperate situation.

To finish off, you are going to look at some basic wilderness survival tips.

Wilderness Survival Tips

The wilderness is a tough place to be for a day, let alone for days, weeks, potentially months on end. Your mental and physical strength will be tested to their limits, and it really is a case of survival of the fittest. Most people think that they could survive in the wilderness easily, but what would you do in the event of an accident? How will you really cope? And are there any tips or tricks that could help you?

Here are six basic steps you should learn, master, and remember if you are to survive in the wilderness:

1. Have the Right State of Mind

In other words, get a grip. Your state of mind is the key to your survival – if you are prone to panic, you won't last long. Panic equals bad decisions, without thinking about the consequences; if you start making impulsive choices, you are minimizing your chances of survival and rescue.

Even if a fast decision is needed, take a minute (or a few moments if the decision is imminent) to relax. You will make much better decisions and find more efficient solutions. While cortisol and adrenaline help you, do some deep breathing to bring your blood pressure down. Inhale for a count of five, exhale for a count of four, slowly. Repeat until you are under control.

2. Always Have a Plan

Start with what is the most important and work from there. For example, if someone has injured themselves, that must be dealt with before anything else – that's where your first aid kit comes in, along with knowledge on dealing with common injuries.

Once that has been dealt with, you and your entire group need to decide who is doing what and how the chores are going to be divided. You also need to decide how long to stay in one place and when to move on.

If you have gotten lost, you should sit tight to make it easier for rescue. If you have a good survival watch, you can get yourself out of trouble – they have compasses, barometers, altimeters, and more on them, ensuring you know what's coming and where to go.

If you opt to move on, you need a plan for shelter, water, and food.

3. Build a Shelter

This is one of the first things you should do when out in the wilderness. If you have your bug-out bag, you should have shelter and a sleeping bag, but if you don't, you're going to have to build one, especially if the night is drawing near.

Find a large fallen log and lift it to lean against a rock or another large tree. That's your shelter foundation. Cover it with branches, leaves, and brush, and if you have one, spread a tarp over the top to keep the rain out.

Clear bugs and sharp rocks from the ground and make a bed from leaves and twigs. If you are caught out in heavy snow, dig a hole and spend the night there – do make sure the entrance can't get blocked by drifting snow or an avalanche, though. Small caves are also ideal shelters, especially as you can build a fire in the entrance. Do make sure no animals inhabit the cave before you move in, though.

Avoid crevices, anywhere that can potentially flood or anywhere that might be home to large wild animals.

4. Make a Fire

This is the next important thing and easily done if you have waterproof matches. Light a bunch of twigs, and keep adding larger branches. Make sure all the wood is dry, or the fire will struggle to light and stay alight.

If you do not have any matches, use friction to create a spark – build a nest from dry leaves and grass, then find a piece of flat wood. Make a small notch in the wood. Put a piece of bark under the notch, and using a pointed twig or small piece of wood, spin it first in the hole, rolling it until embers light the board. Then you can transfer this to the bark and start your fire in your nest.

If the sun is high and strong, and you have glasses, use them to direct the sun at the nest; it will catch fire. The same can be done with a plastic bottle.

5. Find Food and Water

If you don't have your bug-out bag with you, you will need food and water, particularly the latter. If you are carrying a water purifier and have a water source nearby, you have a great source of water. If not, produce your own by collecting rainfall from leaves. You could wrap green branches with plastic bags and wait for them to sweat too.

For food, if you have any with you, ration it. If not, find food. You can set traps or hunt for animals if you have the tools with you, or look for insects, such as larvae, worms, and other small bugs – these are packed with protein, and you'll find them in humid dark places under rocks, and around trees.

Do make sure you know the difference between poisonous and safe insects and plants.

6. Signal for Some Help

To do this, use all that you have on hand three times – a whistle should be blown three times, with a few seconds break, and then again.

Build three campfires; if you are moving, tie ribbons on three trees in a group, or leave three rock mounds.

If you have a satellite phone, you are in a good position to get help by sending an alert message.

The last tip: Whenever you leave for the wilderness, make sure someone knows where you are and what your itinerary is.

Conclusion

Congratulations on making it to the end of this guide. This book should have been useful and provided you with all of the information you need regarding prepping for survival. As you can see, you cannot learn to prep overnight. It is not a hobby; it is a serious business, and one everyone should be learning right now.

It isn't about buying up your entire supermarket and filling fridges, freezers, and cupboards with loads of food, just in case. It isn't about buying up your local camping store, just in case. It is about stockpiling the right stuff in the right quantities. It's about learning new skills that will help you survive in a worst-case scenario. It's about being prepared for whatever happens without being reckless about it.

Why fill your fridge and freezer when the first thing that is likely to go is your electricity supply? Why spend hours making your garden look pretty with flowers when you should be growing food that will keep? Why look at life with an "it will never happen to me" attitude – millions of people the world over have thought that way and millions of people have gotten caught out.

Read and learn because, one day, the S really will HTF – will you be prepared?

Part 2: Homesteading

A Comprehensive Homestead Guide to Self-Sufficiency, Raising Backyard Chickens, and Mini Farming, Including Gardening Tips and Best Practices for Growing Your Own Food

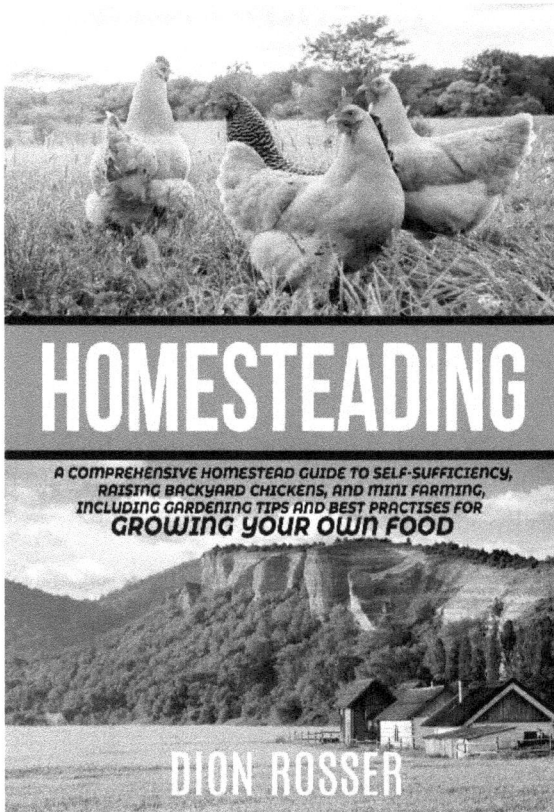

Introduction

What do Academy Award winners Nicole Kidman, Julia Roberts, Reese Witherspoon, and Russell Crowe, alongside *Friends* star Jennifer Aniston and actress Vera Farmiga have in common?

Apart from living life as Hollywood A-listers, these people are also famous for embracing the life of a homesteader. But what is so hot about homesteading that even the red-carpet walkers are drawn to it?

It is ironic that no matter how successful people tend to become and how many different places they go in life, their primal instinct kicks in, somehow, and they always go back to Mother Nature and her bounties. Homesteading has become a catharsis for people who finally want to live the life of a producer rather than a mere consumer.

And if you have been mulling over the idea of raising chickens, goats, pigs, and cattle while also tending a vegetable and fruit garden, then you are not alone. Thousands of people all over the globe have decided to say "yes" to gardening and domesticating animals – and why wouldn't they?

There is so much more to homesteading than just producing your own stocks of milk, eggs, meat, veggies, and even crafting your own soaps and candles. It is also not just putting a parcel of land to good use.

If you are growing tired of a complex lifestyle and just want to live a fuss-free life, then homesteading is for you. It means embracing wholesome values that have been forgotten by many.

Do you want to leave better memories for your children? Then homesteading allows them to conquer themselves – and not just the levels in a video game. If you are stubbornly independent and want to be self-sufficient, then the life of a homesteader is the best way of existence for you.

Homesteading simply makes sense, doesn't it? But, of course, freedom brings with it a plethora of responsibilities – and this is your Day One. Some skills need to be learned (and even unlearned), there are laws to be mindful of, and even many challenges ahead.

The beauty of homesteading, though, is that you change your life for the better – forever. If everyone in your family is ready to dive in, then this is it! Prepare to learn the foundational knowledge and skills that will make you an expert homesteader in no time.

Welcome to your newest adventure.

PART ONE: HOMESTEADING BASICS

Homesteading Explained

Once upon a time, President Abraham Lincoln made available acres upon acres of land to adults who never took arms against the U.S. federal government, women, and immigrants. The Homestead Act of 1862 offered land ownership to these peoples to address the inequalities that were quite rampant back then.

This Act is considered one of the most important legislative acts in American history because public domains – approximately 270 million acres or about ten percent of the United States area – were given to private citizens. Settlers just needed to build a home on their parcel of land, make improvements, and then farm the land for five years so they would be deemed eligible for proving up or becoming the legal owner.

If you look at it, at one point in history, everyone practiced homesteading, where physical labor was the norm. Fast-forward to modern times, and it has turned into a much wider spectrum.

Many people talk about homesteading. Yet when asked to define what it means, there are varied answers. Some say that it means you can provide your own solar, water, or wind-powered electric generators. Some claim that pure homesteading means bartering over the use of money. Still, some define homesteading as letting go of lucrative careers and signing up for a more simplistic way of living.

But how would you define homesteading?

The broadest meaning of this term is being able to commit to a life of self-sufficiency. This could encompass growing as well as preserving your own food, making fabrics for your own clothing and, yes, even providing electricity through natural means.

As you learn more about homesteading, you will realize that not all homesteaders share the same set of values. In fact, it is a diverse bunch of people who say "yes" to this way of life. You will find that there are people who homestead because they are just tired of the daily grind of urban living. Some want to prepare for economic hardships. Still, others just enjoy tilling the land and watching things grow as they plant, harvest, and then preserve their own foods.

Whatever your reasons for considering homesteading or learning more about it, one thing is certain – you have just made a noble decision.

First off, homesteading is a humbling experience, an eye-opener as to how finite life can be and how small you are in contrast to the universe (or multiverse as other people would believe).

Some crops grow and are harvested, while others simply fail. Even the best of farming plans can fail, and so saying "yes" to this way of life is an attempt to tame a piece of Mother Earth and wait for her blessings to come. And as the successes and failures of nature humble you, you also learn to persevere and pray because you have finally comprehended that life is not always within your control.

Homesteading also teaches strength. Talk to older people – especially the ones who grew up on a farm – and they will tell you that they learned to milk cows when they were barely school age. Young girls also cooked side by side with their mothers – all are environments that are now unfamiliar to today's generation.

Kids, back in the day, were capable of thriving even in the harshest circumstances, and this is the very thing that the world badly needs. Homesteading builds the strongest work ethics so that even the youngest children in your family will know how to contribute in terms of labor.

A homestead is one of the greatest places to raise your children. Living in this farm-like setting is schooling in itself. Kids who are trained in these settings can see, hear, and understand more. They will grow up to be more responsible because they are the ones who will think twice before wasting food or polluting their environment.

Talk of growing things and producing your own food, yet another reason for homesteading is the want for food to taste better. You have probably eaten in a restaurant somewhere and complained about the flat-tasting meat or bland salad.

Eggs and meat taste best when fresh, and so do tomatoes and other produce.

The great news is that you do not have to suffer through another flavorless meal anymore. You can now take the conscious effort to live self-sufficiently and enjoy the foods in the process. But to be able to enjoy that fresh tomato, be ready to put in months of hard work as you plant the seeds, make sure that the seedlings grow, and then plan the correct time to plant them in the ground.

And even then, you are not done yet because you need to protect the tomato plant against the harsh weather, set up a trellis for them, and then wait for the right moment when they have ripened.

At the end of the day, you will also realize that homesteading instills appreciation in you because nothing better teaches about this than hard work. You will also have a renewed sense of value for growing your own nutritious food. Moreover, the amount of energy that you need to exert just to raise chickens, pigs, or cattle is astonishing. Add to this the thought of butchering the animal that you raised, and the work becomes doubly difficult.

No matter how difficult, though, all these add to your knowledge of Mother Nature and how learning about gardening, breeding animals, bee farming, pollination, making cheese, and such, could lead to true freedom and independence.

Now, don't think that you are going to be like your great-grandparents who had to plow the land by hand (though this is your choice, too). The great thing about the 21st century is that you have access to farming equipment, better health care, and the means of staying connected to the world (some homesteaders blog and keep Instagram accounts so they can inspire others). All these can be used for and not against your chosen lifestyle as a homesteader.

Homesteading is also far from an easy life because you need to commit to physical, emotional, and mental labor to make this lifestyle a success. Oddly, even when it requires these different human aspects, it guarantees that you will live a stress-free life.

So how does homesteading equate to zero stress if manual labor is involved? The answer is in the amount of security and peace that is often linked to self-sufficiency and so, in turn, to homesteading. These days, people have the chance to live a healthier lifestyle when they choose to live in home-based farms or even when they simply decide to grow an indoor garden.

Many homesteaders have reported that living on a farm or in the country is much more relaxing than setting up home in a city. They also feel freer and healthier in such an environment. They can even save money on clothing and food. Some even make money through trading and selling the goods that they produce in their homes.

And who could argue the wellness that comes with the great outdoors? Even young kids can learn about their connection to nature. City life only provides parks, but homesteading offers much more. It offers respite from the serious threats of genetically modified crops and foods.

So, what is this big fuss over genetically modified foods?

It has been 27 years since the first genetically modified food was introduced onto the market, and the American diet has since consumed huge amounts of this manipulated foodstuff. It is not surprising that genetically modified organisms (GMOs) or genetically engineered (GE) products now dominate many agribusiness sectors. It is likewise alarming that biotech and chemical companies now control seed suppliers.

You are eating these dangerous foods if you are not buying organic or not growing your own food in your backyard. Just think about the modified crops that are now in many grocery shops – corns, rapeseed (used in making canola oil), peppers, zucchini, squash, sugar cane, rice, papaya, and peas – and you would be distressed about where to get your next nutritious meal.

The GMO producers are literally out of control, insomuch that the small print will not even indicate that the corn and soy that you bought are GMOs. Prevention Magazine cited that 80 percent of processed foods are genetically modified – and that is a lot. So, it matters that you check if your food has a non-GMO label. What is even better is if you grow your food, that way, you are guaranteed that no pesticides were used.

Choosing to become a homesteader, therefore, becomes much more promising because more people are joining the bandwagon. In a world of genetic engineering and GMO companies winning court battles against farmers, the good news is that homesteading is now a global social cause.

This is the perfect moment to turn into a homesteader and not worry about the high expenses brought about by living in the city. And as the job market is becoming more erratic, there is the logic in considering country living and joining many others who now enjoy healthier, more peaceful lives.

Modern homesteading offers more possibilities and choices nowadays, so roll up those sleeves because it is time to sing heigh-ho.

What Type of Homesteader Are You?

Now that you know the benefits that come with homesteading, it is time to find out what kind of homesteader you are. If Dictionary.com defines the term as any dwelling with its land and buildings where a family makes its home, then the act becomes a lot simpler but, nowadays, homesteading goes beyond setting up a structure and letting your family stay there.

Since homesteading is also about being self-sustaining, then it could also mean that it is not just another place where you feel at home; it's also the zone where dreams are built, and wellness is upheld.

Homesteading is the place where the family lives as one and works as one. Sure, it could mean living off-grid for some families or just wanting to save some more money on organic foods, but ultimately, every family needs to set their homesteading goals.

Homesteading could mean many different things, but one of its most important goals is self-sustenance or being less dependent upon others. There are also five different kinds of homesteaders, and it is time to find out which of these you are.

The Urban/Apartment Homesteader

One of the greatest things about modern homesteading is its flexibility. Gone are the days when homesteading is thought of as a thing for ranchers and farmers only. There is no need to own thousands of acres of land anymore just to be called a homesteader. These definitions are quickly changing so, no matter where you are, you can now be a homesteader.

If you are currently living in the suburbia, specifically in an apartment, no worries, you still have unlimited possibilities in terms of starting your homestead.

But what would homesteading look like inside a tiny apartment? How about in a small backyard?

Living in a city should not limit your homesteading dreams. You could live in the heart of the city jungle and have breakfast on a tiny balcony each morning but could still live the life of a homesteader.

Characteristics that Define the Best Apartment Homesteaders

• The dwelling unit is small. Homegrown fruits and vegetables are your keys to sustainable living, but where there's limited space, these plants have no room to grow, so if you've been wracking your brain on where to put your vegetable pots, then you're definitely an urban homesteader. The urban homestead is also characterized by single-family habitats.

• The household size is about three persons. A report prepared by the United States Department of Housing and Urban Development Office of Policy Development and Research showed that the mean number of persons in an urban homestead is at three point two.

• A newly formed household. The same study also proved that about 15 percent of the urban homesteaders said they were not heads of their previous households.

- Previously lived in rental properties. The statistics show that 90 percent of urban homesteaders were previously renters.

- Full-time or part-time employees. It's also been found that 90 percent of urban homesteaders were working full time or part time before becoming homesteaders.

- Modest savings. One of the probable reasons why urban dwellers also want to become homesteaders is the chance to augment their incomes through gardening or keeping livestock.

- Optimized use of space. First, you can grow different herbs in small pots, but you have to know the right spot to put them. And the best apartment homesteaders use areas such as the windowsill, the balcony, and even nooks and crannies just to grow their veggies and herbs. These are the people who enjoy tending onions, oregano, garlic, and herbs in smaller containers. If you're okay with such a setup, then you probably live in a tiny apartment at the moment. If the idea of a hydroponic garden is not just exciting to you but also a solution to your gardening issues, then you're definitely an apartment homesteader. Likewise if the idea of visiting local farmers at local farms excites you. When you're free, and your idea of "me time" is to visit the farmers' market to buy fresh milk and eggs, then it's possible you don't currently enjoy them because you're living in the city – specifically, in a small space.

- You want to begin a homestead pantry for your family. So what are the skills that city-dwellers turning into homesteaders often have? Small-scale fermenting, canning, seed saving, knitting, sewing, water management, rooftop beekeeping, and even soap and candle making are skills that were common to your grandparents. But these are now common urban homesteading skills, too.

If you are also in search of neighborhood elders who can teach you the skills or attend lectures and classes in your local area, then you're a true-blue urban homesteader. This is also true if your source of knowledge is the city library.

If you have been secretly wishing to learn food preservation so you can preserve your bulk food purchases, then you are ready to become an urban homesteader to provide food for your family.

On a personal level, if you prefer wearing barefoot sandals while going around and spraying the organic anti-bug mixture on your potted plants, or if your idea of a pet is having a cat, a toy dog, and a cute turtle, then these just prove that you are an urban-dweller.

Don't worry, though, because even if your idea of a vacation is to visit city museums, you could still start homesteading. It is all a matter of learning the right skills meant for such settings.

The Medium to Large-Scale Homesteaders

What makes bigger scale homesteading different are aspects such as fruit trees, a berry patch or two, a vegetable garden, bigger composting, rain barrels, livestock, and bigger parcels of land in contrast to the limited spaces in the city.

The land size also indicates whether you are a small, medium, or large-scale homesteader. The measurements are:

Small homestead – six to ten acres of land.

Medium homestead – 11 to 30 acres

Large homestead – 75 to 200 acres

Apart from the acreage, here are the telltale signs that define larger land homesteaders:

• Hard work is your everyday reality. Your idea of a workout is hiking, kayaking, lifting heavy objects, or taking long walks around your neighborhood.

And if you don't even need a cup of coffee to jump-start your morning, then you're probably already used to routine, morning physical work and are, therefore, meant for larger-scale homesteading.

• You dream of rows upon rows of different plants. If you can't picture yourself tending to mere pots and containers, then you're definitely cut out for large-scale homesteading.

• You are thinking of expanding the workforce. One of the most obvious signs that you're about to live in a large homestead is when you're already thinking of employing more people to help you with your homesteading operation. If your family members are no longer enough to cover all of the tasks on your homestead, then you know you've got a larger piece of land.

• You're also thinking of bigger investments. Bigger workforce, livestock, and gardening all require additional money or investment. Most of the time, the glaring large-scale homesteaders add bigger machinery and equipment so that the work on the homestead is made lighter.

Machinery That is Often Added

o Tractor (costs north of $10,000)

o An all-terrain vehicle (small ATVs can cost as little as $1,000 while the biggest ones could cost as much as a tractor)

o Farm truck

o Wagon

o Backhoe

o Cultivator

o Plows

o Harrows

o Balers

o Harvester

• You think you need an irrigation system. You know that you're a medium to large-scale homesteader if you've been planning to set up an irrigation system to water your garden.

• You want to share your produce with your neighbors. If you desire to grow food beyond what you can store in your family's homestead pantry, then you're ready for a large-scale homestead. You are more than a hobby farmer at this point; you're already thinking of blessing lives outside that of your home.

So, have you figured out what kind of a homesteader you are? Now it is time to answer a simple "yes" or "no" to a series of questions:

1. Do you often worry where your food originates?

2. Do you wish to become self-reliant?

3. Do you like cooking your own food, or are you willing to learn?

4. Are you okay with learning traditional skills, such as carpentry, sewing, knitting, candle making, doing your own laundry, and others?

5. Would you be willing to learn skills so you can protect your family during an emergency?

6. Could you butcher animals for organic meat?

7. Would you consider the thought of raising animals for your sustenance?

8. In your opinion, is living off-grid an appealing way to live?

9. Are planting and food preservation exciting things to do on your own or with your family?

10. Would you be able to live miles away from the grocery store?

If you answered "yes" to most or all of these questions, then you are absolutely ready to wear those farming boots.

Steps Before You Start Homesteading

Congratulations! At this point, you have bravely said "yes" to becoming a homesteader. Now your sleeves are rolled up, and you are ready for the tasks, so what is next?

Your next step is to prepare mentally and physically for the days and years of hard work ahead. This phase is the key to staying and not giving up once the challenges appear.

Planning for homesteading is not all about knowing what to feed the chickens or how many gallons of milk can be squeezed from one cow. Before you even buy your first packet of seeds or your farming boots, you need to know some behind-the-scenes stuff so that you won't be caught unaware.

Being in control of your food, health, and home, in general, is commendable.

Here are the crucial steps that every future homesteader needs to take before making the huge leap:

Commit

Starting a homestead from scratch is a daunting task, so it helps to break the responsibilities down into smaller and easier steps. Decide the things that you want to commit to before you even set foot towards achieving them.

Take the time to research, read, ask the right questions, and take down notes. Be ready for the unexpected by setting up buffers (e.g., knowledge and money).

Another part of your commitment is to alter your mindset. As soon as you are ready, decide to make things rather than readily buy them, or reuse rather than toss them. As you start preferring these, then you are on your way to becoming a homesteader.

Your life is no longer a life of consumerism, so a part of the commitment is also to downsize. A coffee or waffle maker will no longer have use in a homestead. There are many items that you currently own that should go. The things that have been in storage bins and boxes for years are probably not going to be used anyway, so donate them, give them away to friends, sell them, or simply dump them.

Assess Yourself

Statistics show that there are more introverted homesteaders, and this is easy to understand because the wide-open spaces that typically come with homesteading foster much alone time, fewer neighbors, and room for the kids to run around.

Homesteading is synonymous with adventure once you set out to do it because the clucking chickens, growing vegetables and fruits in the garden, even the pantry that is slowly filling up with preserved food in jars are as exciting as homesteading could get. But there is a side to this life that could get flat-out melancholic as living in the country potentially leads to isolation.

As soon as the dust begins to settle (so to speak), you might begin to miss the conveniences that are offered in the city – from ready-made meals to the cinema. The walls could feel confining if you have not taken the time to assess yourself before homesteading. If you are an introvert, then you could get energized when you are alone as compared to the extroverts who get their energy by socializing with people (now this second batch will need more help in terms of keeping sadness at bay).

This does not mean that extroverts are not effective homesteaders, though. It is just a reality that a few families have struggled with. Rural loneliness is real, as people have admitted during their homesteading battles, but there is much room for different types of personalities when it comes to the homesteading gig.

Understand what motivates you. Introverts are project-driven so you can focus your energies on the projects that are in front of you. You have probably heard of homesteading mothers who took care of their kids while they also deep-cleaned their house and gardened about but who still felt bored, so think twice if you can take this project head-on.

It also helps to understand that rural living does not mean being cut off from the rest of the world. You can still access social networks so you can join forums where you can share ideas with like-minded people. This could be a great source of social stimulation. Even when these people are just your online friends, still, you get to keep a social discourse and, in the process, maintain your mental balance.

Assess now if you can keep yourself occupied and not be dragged down by the so-called cabin fever (irritability and restlessness as a reaction to claustrophobia). Homesteading requires a 180-degree turn from an easy life, so face the mirror and be honest with your answer as you ask yourself: Are you truly ready to become a homesteader?

If the thought of being self-sufficient and knowing where your food comes from weighs more than any difficulties ahead, then go ahead and sign up to a life of homemade ice creams and homegrown meats.

Find New Friends

You can socialize online, or you can do it the old-fashioned way as you find homesteaders like you (preferably your neighbors) and befriend them. You will even have more people to barter with for your future produce and livestock.

And those homesteading questions? You can finally put them to rest if you ask your new homesteading friends about their experiences of homesteading laws, weather, general climate, etc.

Having friends also means having someone to borrow things from during emergencies. You can borrow a dehydrator from a new friend in exchange for your canner.

Plan

There is a saying that if you fail to plan, then you plan to fail. As with anything, planning involves setting realistic goals.

Many people feel overwhelmed when they start homesteading. This is understandable, but it can be managed when you set goals that you can reasonably handle. You can set seasonal goals instead of distributing your tasks across different goals. This will just lead to confusion, so tackle your projects one month or one weekend at a time.

You also need to plan on an extra income while you are still establishing your homestead. It is okay to dream that homesteading can provide all of your needs, but this is not realistic.

You need to seriously consider the expenses that come with setting up your self-sustaining habitat. Also, you might be used to eating out or traveling, so you need to find out how much of this you are willing to give up.

What is great about modern homesteading is that your lifestyle choices have greatly widened, as more income opportunities are now available to you. As opposed to your forebearers, you now have access to more information, so you are less likely to commit the greenhorn homesteader mistakes.

This could include taking on an online job while you are homesteading. Thanks to the Internet, there are now more job opportunities, even in remote areas.

An online business is also not that expensive to set up. Selling wood cookstoves over the Internet might not be a lucrative source of income, but it can still provide a few thousand dollars annually, enough to cover some of the homesteading expenses.

Write Down Your Plan

You can also brainstorm as a family regarding the homestead layout. A graphing notebook is your best friend as you sketch the floor plans to scale. Fill the pages with storage space, the pantry, and other functional spaces.

However, the layout is not the biggest concern in terms of planning for homesteading – financial preparation is. But don't worry – there are basic money principles that can help you cope.

First, say "no" to debt. Starting your homesteading project using borrowed money is not a good idea. This goes against every homesteading principle, more specifically, self-sufficiency.

Homesteaders generally want to detach from money issues. An experienced homesteader prefers bartering rather than buying anything. Many just leave a single credit card for emergency purposes.

Homesteading also means being prepared for emergencies, such as equipment damages, injuries, illness, and such. Be sure to set aside an emergency fund.

Keeping up with the Joneses is not the way to live your homesteading life, especially when you want to avoid bankruptcy like the plague. Just stop comparing yourself with other Pinterest, picture-perfect homesteading families. Remember that what you see on people's social accounts is just the façade and not the real picture, so just water the grass from your side of the lawn, and you will be awed at how your mindset improves.

Financial preparation also means being aware of money pits, especially when your budget is tight. You will soon grasp that generating your own food is pricier than buying it at the store, and it is much easier to purchase a gallon of milk than to keep a cow for milk.

However, homesteading is now your way of life; it is raising kids who appreciate Mother Nature, the value of hard work, and many other things. So, yes, homesteading – at least some aspects of it – is going to cost more.

Have an abundance mindset – you need one if you want to transition from an easy life to homesteading successfully. With this kind of mindset, you will feel more at peace about time management, resource management, and even your finances.

Embrace Minimalism

While you are encouraged to have a positive mental outlook, this does not mean keeping a garish lifestyle. It can be so easy to get caught up with the thought that you always need to acquire more or do more in life, but doing only the things that matter is the best way to go.

If there is anything in your life – anything at all – that has been draining your energy, time, and money that you can finally eliminate, then now is the time to let it go.

If your kids have signed up for a lot of activities at school or if you have had to drag yourself to an event like book club, then you just have to say "no." Even if these things are good, if they don't make you truly happy, then there is no point in adding them to your calendar.

Look at Challenges as a Way to Foster Growth

Homesteading is a way of life that can be demanding, but it can also be fun and rewarding. Setting your mind to this will help you through the roughest patches. Understand that every setback and mistake will provide a learning experience.

And don't forget to ask any farmer to tell you about his or her past errors and you will surely be told that those mistakes also helped stage their eventual success.

Find out if homesteading is right for you. If you like providing for you and your family's needs, and putting in hours of physical labor won't deter you, then homesteading is something that you will surely enjoy.

11 Essential Homesteading Skills

Okay, homesteading means survival; you get that now. Yet even the most efficient carpenter, welder, and truck driver can have the skills that will work fantastically on a homestead, but, individually, they might not survive. You need to have an information arsenal that will cover topics that are needed in sustaining your new chosen lifestyle.

However, you do not have to become an expert gardener on your first day. Knowing how to plant and cultivate are skills that could help you, though. Having a passion for them will even contribute to a livelier garden.

You can start small when it comes to your skills, and today that is what you are going to do. Begin small by taking the time to know which homesteading skills you need to tackle first.

Composting

Every time you repurpose products or reduce waste on your homestead is a win-win situation for you. Remember that composting is the skill that conquers both.

The compost area is not difficult to set up whether you are working on a small piece of land or a bigger one. You can take the easier route by using a trash bin as your initial compost area or install a more complex compost area using palettes or plywood.

Composting skills can be as basic as gathering grass clippings, straw, natural food waste, and other organic materials, at first. When this pile of waste decomposes, then you already have a compost soil that can be used as a garden fertilizer.

Gardening

Are you planting the correct kinds of crops for the climate in your area? Are you also providing ample space for the root vegetables to grow? Will you be able to distinguish and prevent crop diseases?

Study the crops that you will plant and then strategize your planting methods. Till the ground, use the windowsills, hang pots, create a fence of edible plants, and use just about every possible piece of the landscape on your homestead to maximize the use of your land. If you live in an apartment, make the most of every available space.

Study which plants grow in what parts and which ones can grow side by side.

In a nutshell, your gardening skills should include knowing:

- When to plant.
- The distance between planted seeds.
- What requires propagating.
- What should be included in a greenhouse.
- What pot sizes to use.
- Germination and maturation periods.
- What pests will ravage your crops.
- What vegetables work together.
- When to harvest your crops.

Salvaging

Your fruits and vegetables are all great sustainable sources of organic food. So, the next skill that you need to learn is how to save the seeds that you can replant properly. Bell peppers, cherry tomatoes, carrots, and blueberries can be replanted. All you need are a sharp knife and pruning shears to harvest the seeds and then store.

Using the pruning shears, cut off the seed pods. Place the pods in a room with a warm temperature. This has to be a low-humidity room. Spread these on newspaper or several pieces of paper towels.

Let dry for one-two weeks.

Food Storage

Any city dweller can always run to the local grocery store to buy pasta, flour, and canned goods, but these will not provide you with a balanced diet. You still need fruits, protein sources, and vegetables to obtain essential vitamins and minerals – and what better way to flood your system with these essentials than to eat organic?

There are many forms of food storage, and you will have to learn these to live on your homestead effectively. You can do any of these inexpensively using techniques such as canning, drying, and freezing.

Canning is the application of heat to food that has been sealed in jars. This process destroys microorganisms that lead to spoilage. The right canning techniques prevent spoilage by food heating for a specified time and driving out air from the jars, thus, creating a vacuum seal.

The United States of Agriculture (USDA) approves only two canning methods: pressure canning and water-bath canning.

Pressure canning uses a big kettle producing steam inside a sealed compartment. The packed jars that are in the kettle need to reach an internal temperature of 240 degrees. A dial gauge is then used to measure the amount of pressure that the jars go through.

Pressure canning is best used in processing meat, fish, poultry, and vegetables.

Water-bath canning, on the other hand, is the method that is also called hot water canning. It also uses a big kettle with boiling water.

The difference is that the filled jars are placed underwater and then heated up to 212 degrees for some time. This method is best used for foods that have a high acidity level, such as fruits, tomatoes, and pickled foods.

Of all the food preservation methods, drying is the oldest. A portion of the food that is being dried is exposed to high temperatures so that moisture is removed but only to a point where it will not cook. Air circulation is a factor in evenly dried food.

There are now electric dehydrators complete with a fan and a thermostat that help regulate the temperature for food drying. The oven can also be used, or the heat of the sun, but these latter processes take longer, and the results are far inferior compared to those that were dried inside a dehydrator.

Food freezing is a process of preparing, packing, and subjecting foods to freezing temperatures while they are still fresh. Fruits, vegetables, fish, meats, cakes, pieces of bread, casseroles, and soups can all be frozen.

To freeze food, be sure to wrap food tightly and to keep the freezer door closed at all times. Plan the foods that you will freeze so that the unit will maintain its ideal air circulation.

These processes, along with other food preservation methods, such as fermentation, alcohol immersion, pickling, immersion in olive oil, and preserving in sugar and salt, are time-tested survival skills that your predecessors learned and effectively applied.

Milling

Depending on where you chose to homestead, storage concerns for larger quantities of grains and flours could pose a challenge. You can still grind a few pounds of grains at a time and then have it stored for baking and cooking purposes.

The simplest modern way to mill flour is to use a food processor. Feed ten cups of flour into the hopper; then, you will get ultra-fine flour in a few minutes. If you want to do it naturally, then you can use the good old-fashioned mortar and pestle. You can also buy a mixer attachment specially made for milling grains or a portable grain mill.

Butter and Cheese Making

There is no need to own a cow just to make your first homemade butter. What you need is just a bit of cream and a jar that you can shake it in. As for cheeses, you can also make ricotta cheese, mascarpone, Chèvre, and maple yogurt cheese.

Cheeses can be made from any kind of milk, including sheep, goat, and cow. The kind of coagulant used depends on the kind of cheese that you want to make. Acid cheeses need an acid source like acetic acid (vinegar acid) or a glucono delta-lactone or mild food acid.

Method:

1. Begin with fresh, warm milk. The fresher the milk, the more scrumptious the cheese.

2. Acidify the milk using citric acid. A liter of milk should be mixed with three point five grams of acid.

3. Add a coagulant (rennet).

4. Using your clean hand, press the surface of the milk to test the firmness of the formed gel.

5. Using a cheese harp, whisk or knife, cut the formed curds. Keep in mind that the smaller the slices, the drier they will be. Cut bigger pieces, and you'll have a moister cheese.

6. Turn on the stove, then continuously stir while cooking the curds.

7. Wash the curds (whey removal) by taking some of the whey off the vat and then having it replaced with water.

8. Drain by using a colander. You could also wait ten minutes so that the curds will finally settle to the bottom.

9. Press the curds at the bottom while they are still hot.

10. Salt and age.

Animal Husbandry

The most efficient homesteaders are those who depend on animals. They are those who have horses to ride, dogs that guard, chickens that provide meat and eggs, and pigs that eventually become bacon and ham.

The Encyclopedia Britannica Online defines animal husbandry as controlled cultivation, management, and production of domestic animals, including the improvement of the qualities considered desirable by humans.

Animal husbandry is also associated with beekeeping, dairy farming, and breeding – you will learn more about this later.

Starting Fire

The cavemen have done it so you can do it, too. As a survivalist, building a fire should already be second nature to you. The drill is this:

1. First, create a tinder nest. This is where the starter flame will be sparked.

2. Next, make the notch and place bark under it. Begin spinning, then build your fire. Prepare the fireboard and then keep rubbing.

3. Start the fire.

If you feel like taking an easier route, then use a magnifying lens to start the fire. Just put the tinder underneath the lens and wait for the beam of the sun to produce a flame.

You will need this skill to cook your food when there's no electricity. The classic bonfire will not just provide warmth but also a place where you can cook your food.

You can also still rely on propane tanks paired with a reliable gas stove. You can also just grill your food to make even more scrumptious meals.

Cooking

Just as important as growing your own food is learning how to cook healthy meals. If you have been relying on restaurant meals, don't fret, because this skill can be learned.

It is crucial to shop for your own cooking utensils. The essentials are wooden spoons, stirring spoons, spatulas, measuring cups, and a whisk. You should also invest in pots and pans (copper would be a nice material for these). Use cast iron as much as you can as it makes any food taste better, plus it is a durable cooking tool that you can use for many years.

One good paring knife should be enough to begin your cooking lessons.

Purchase cookbooks that were written for beginners. Take the time to read the reviews about each cookbook. If you do not have tangible copies of cookbooks, then you can search for phone apps.

DIY Natural Remedy and First Aid

Homesteading almost always equates to living miles away from the nearest pharmacy or hospital, so you need to be prepared for self-healing. Concoct homeopathic remedies for minor conditions.

Find herbs and spices in your garden that can be turned into remedies for common ailments. You can make homemade soaps, lotions, and salves. You can also harvest raw honey that is known for treating skin rashes, colds, coughs, burns, and many more ailments.

Another useful set of skills is tending to cuts, burns, and scrapes to prevent infection. Keep a stock of medicine suture accessories, and even a military-grade tourniquet.

Learn also the Heimlich maneuver as well as CPR. Remember, though, that the person still needs medical help after the performance of either First Aid. To do the Heimlich maneuver, do the following:

1. Get behind the person.

2. Lean him or her forward and then give five back blows using the heel of your hand.

3. Put your arms around the choking person's waist.

4. Make a fist, then put it above the navel with your thumb in.

5. Wrap the fist with your other hand, then push upward and inward.

6. Do five abdominal thrusts.

7. Repeat until the person can expel the lodged food or cough on their own.

As for CPR

1. Put the person on a steady surface.

2. Check the insides of his or her mouth, then tilt their head and chin.

3. Do 100-120 chest compressions every minute.

4. Add two rescue breaths for every 30 chest compressions.

5. Do this until medical help arrives.

Fortification

You also need to fortify your homestead against robbers and intruders. The ever-reliable security cameras, motion sensors, and traditional alarms play a huge role in keeping your family safe and secure.

There are several options for fencing, including:

- Post and board fence (often seen around horse pastures)
- Split-rail fence
- Electric fence
- Woven wire fence

A woven fence is a good way to secure your homestead because this is a sturdy boundary. You can even top this with barbed wire to make it even more secure.

Your farm dogs can also become your guard dogs. They are one of the best deterrents against robbers.

There are many other homesteading skills that you can add to this list so that you can make your homesteading adventure even more exciting.

How Much Does Homesteading Cost?

The first time that you delve into it, homesteading can seem simple and quaint. After all, what can be more idyllic than canning your foods, your kids playing with farm animals, or everyone frolicking under the sun, right?

However, as soon as you stake the first post of your fence, you will realize that homesteading can be pricey – especially when you are living off of a single income. Yes, remodeling the house, improving the land, putting up outbuildings, and growing animal herds all cost money.

There are two ways this could go – starting small or going full blast the first time. Say you want to begin homesteading less than an acre of land, where you will have a little house, a garage for two cars, a studio, greenhouse, chicken enclosure, berry bushes, apple trees, a fire pit, and a compost pile – how much do you think it would cost for the upkeep?

The greenhouse could easily cost $1,500 if it is fancier than the $500 pop-up varieties. Other approximate garden costs are:

- $120 for the sod cutter
- $120/day rental for the Rototiller
- $50 for the garden tools (although these can be borrowed from your friends or family on your first few days of homesteading)
- $20 for the seed starting trays
- $160 for the potting soil
- $28/bag of organic worm fertilizers
- $200 transplants
- $25 for the mulch
- $40 for every apple tree
- $8 for every berry bush
- $1,000 for the fence
- $20 for the chicken wire
- $100 for various seeds
- $20 for monthly watering

You can also construct tomato cages using scrap woods. The approximate amount that you would spend for a 1,300 square-foot garden would be about $3,600 from the outset.

This is just a rough estimate of what you would purchase because you would eventually add other products to your monthly expenses. You can cut down on some of the expenses, such as the worm fertilizes and some of the seeds when you finally learn to make your own fertilizers and salvage the seeds.

Transplants can be cut down to 50 percent or be eradicated from the list.

If you are planning on taking care of chickens, here are the costs involved:

- $12 per chick
- $70 for the brood tub

- $12 for the heat lamp
- $15 for the chick feeder
- $15 for the chick waterer
- $200 for the construction of coop and run
- $26 for the hen feeder
- $32 for the hen waterer
- $50 for the water heater (to be used during winter)
- $240/year for the feed or about $20 per month
- $228 annually for the treats
- $60 a year for the bedding

Keeping six chickens would be about $960 every year with an annual upkeep of $500+ for the feed, treats, and bedding.

Beyond these initial amounts, the reality of homesteading is that you can realize your dreams and live the life of a retiree way before you reach your golden years – and these are priceless.

You do not need huge dollar amounts in your bank or investments just to jump-start your homesteading journey. And now that you are a homesteader think of your investments as not just being in the form of money anymore. Even your time can be invested now. An example is when you begin setting up your garden for your self-sufficiency project.

It will cost money to buy the first seeds and other gardening tools, but a huge portion of your effort will result in stock vegetables and fruits in jars and cans. These can last for weeks – even months – depending on how you ration them.

With these foods and eventually the meats to be provided by the livestock, you can make more investments. You will need to keep challenging yourself if you want to live off, or semi-retire on, your homestead.

There isn't a single correct way to do it, but there are guidelines. At the top of these rules is to leave your comfort zone. Pick a monthly expense and then find the average each month for the last few years. You can then think of ways to lower that expense or eliminate it. This will relate to savings that you can also replicate with other expenses.

Homesteading is slowly setting yourself free from the usual monthly expenses. Most people have the following expenses to worry about each month:

- Mortgage or rent
- Insurance
- Taxes
- Gas
- Car payments
- Utilities
- Clothing, and
- Food

Rent or mortgage can be totally eradicated by buying a mobile home (for small-scale homesteaders) or purchasing a parcel of land (for large-scale homesteading). The former could cost as little as $5,000.

Property insurance will cost less when the value of the land is less. Total ownership of the assets will even free you from further expenses. You can also buy a used car instead of a brand-new one so you can pay for it in cash or set up easier monthly payments.

Think of the type of vehicles that you will need on your homestead. Keep in mind that you will need a pickup truck for the farm chores and a car with decent mileage. Older vehicles will greatly reduce tax and insurance costs. Many states have personal property tax assessments on vehicles annually. If you are not going to commute from your homestead to work, then you can save money on gas or diesel fuel.

As for the phone bills, find out which cell phone service provider will give you the most competitive plan. There are now Internet phones costing only $19.95 annually (minus Internet access).

This kind of low-cost phone or cell phone is most suitable for you if you need to save some more money. Check if there is rural Internet access in the area where you are moving. You might resort to satellite Internet if the cable, DSL, and telephone companies won't come out too far.

The costs are also still competitive, especially for the wireless offers for laptops. As for the electricity, you have the choice to use solar panels and other power sources, such as wind and water.

You can have a private well (or more) on your homestead. This could be a literal hole in the ground or one that is run by a good pump. You could install water filters so that your drinking water remains potable.

As of 2020, the cost for good drilling is $5,500 for a 150-foot depth. The price per project could vary, but the range could be anywhere from $1,500 up to $12,000. Residential water wells cost $15 to $25 per square foot, while agricultural or irrigation systems cost higher at $35 to $55 per square foot. An aquifer in place of an electric pump is estimated to be $25 to $45 per square foot.

If there is an existing well on your homestead, but you cannot fetch water from it anymore, then you could drill it again. The cost for re-drilling is $300 to $600 for the labor plus materials.

Growing your food is the most common motivation for many people to choose a life of homesteading. However, you have to consider gas expenses if you need to transport the harvested vegetables and fruits to places where you can sell or barter them.

Crafting also costs money, depending on what sort of crafting you will take on. Your options include hobby crafts such as crocheting, quilting, and knitting, or crafts that can add to your homesteading income, such as soap and candle making, paper crafts, essential oils production, and even sewing curtains.

Large or small, chickens cost less to feed when they roam freely within your protected area. The eggs that they lay can even be added to your items for bartering or kept for your consumption. There is less to zero fuss with these free-range chickens.

You can also raise rabbits for their meat. This is another affordable, healthy way to eat and provide for your family. For good quality hay, pellet, and fresh foods for your rabbit, an estimate is about $25 each month.

Raising dairy cows (costs between $2,000 and $5,000 per cow) and goats (price range is from $75 to $300) also reduce your reliance on commercial food vendors. With these animals, you can produce your own milk, ice cream, cheese, and butter. If you have more of them, you can even barter or sell so that you can purchase other goods with money.

And where there are livestock, fruits, and vegetables, there is also trash. Many homesteaders take their trash to transfer stations for free. For the less understanding local governments, there could be a minimal cost for trash pickup. The monthly trash collection cost can be $15 or so. This is an added convenience to your homesteading lifestyle, but you could choose differently by keeping a compost pit and learning to recycle.

One other major purchase that you need to prepare for so that you can formally begin your life as a homesteader is how to buy the perfect piece of land for homesteading.

You will learn this in a separate chapter.

How to Avoid the BIGGEST Homesteading Mistakes

Okay, so you are ready to embrace the whole new level of freedom that homesteading offers. Those who have started before you know that it can be a trial and error process, but the more that you learn the basics, the better off you will be. And what better way is there to learn than to steer clear of the mistakes that were committed by other homesteaders?

The Most Common Homesteading Errors

Insufficient Research

Homesteading can look easy via its façade. You purchase chickens, place them in coops, and then you wish to collect their eggs and, eventually, their meats – but all this entails effort and time.

Be sure to do your research – lots of it – and interview homesteaders who have done it for years.

This book should provide the fundamentals of homesteading, so take the time to read and apply what you learn.

Zero Planning

Since one of the necessary steps of any project is to start with a plan, not having one will definitely result in a disaster. Whether you just want to have a small yard or wide acres, it is crucial to begin with a plan.

Imagine not having a plan when you are faced with these issues:

• The number of animals that you can humanely and safely raise

• The cost of livestock farming

• The plants that grow best in your locality

• When to plant, etc.

You also need to set up a business plan before moving to the country. As you buy your acres of land, what would you support your family with? How will you pay the mortgage?

Will you sell your animals, or will you become a homesteading consultant? Are you going to be a homesteading blogger?

Whatever you plan to do, set up more than one thing because you need some form of income to support your new homestead.

Expanding Too Soon

Making unrealistic plans is, technically, the opposite of not making plans at all. However, the downside to this, of course, is that people who have unrealistic plans do not even realize that they are impractical.

Idealism is great but not to the point of unfeasibility. You need to seriously assess if the goals that were set are really reachable. Also, consider if the timeframe that you set is even possible.

For instance, raising some chickens is realistic, but doing this on top of building a fence, building the coop, digging a well, and clearing the trees, all during the first week, just won't happen (unless you have additional human resources). This is the perfect recipe for burning out long before you even begin homesteading.

Break the goals down into smaller, more realistic tasks.

Setting the Wrong Budget

There are some homesteaders who, in their excitement, must have calculated their homesteading budget incorrectly. The common problems that could arise from setting the wrong budget include:

● Not being able to buy enough seeds during planting season.

● Inability to cover expenses brought about by emergencies (e.g., human or livestock ailments and machinery and equipment repairs).

● Running short on income, especially when you decided to quit your job when you moved.

Budgeting ensures that you will be financially secure and free from debt (or, at the very least, have minimal debt) until you can fully depend on your homestead.

Since you will be learning about homesteading income sources later on, what you need to do, initially, is to create a monthly budget. A simple notebook and pen can be your tools, or you can use an Excel sheet or go on the Internet and look for apps that you can use to jump-start your budget planning.

Remember that every monthly budget should include the following:

● Recurring bills

● Upcoming/irregular bills (e.g., quarterly or annual bills)

● Seasonal or large purchases

● Loans, debt, mortgage, car payments, etc.

Set up the spreadsheet or make a chart in a notebook. Some people just use tables on a Word document.

Next, add all the income streams and deduct the monthly bills. Be sure to indicate the due dates so that you will not miss them. Be sure to set aside savings even when you are still paying off some debts.

Insert the formula on your Excel sheet or manually calculate the figures. Collect receipts so it will be easier to compute your expenses.

Never rely on your memory alone. It is better to keep records or take a photo using your phone camera each time that you make a purchase.

Recordkeeping is also not limited to accounting for your income and expenses.

You can keep records of the crops that you recently planted, livestock, harvested crops, and how much you were able to preserve.

Think of running your homestead as you would a business – you need to be on top of it no matter what. Never approach homesteading as a mere hobby because farming is a serious business. Every new project that is added to your homestead is, in essence, an investment.

Zero Land Inspection

The problem with not evaluating the land that you are going to use for homesteading is that it might not produce the yield that you initially expected. Soil management is going to be a part of homesteading, and so you would want to physically inspect the land that you are going to invest in. The land, after all, is going to be the foundation of your upcoming homestead.

Take the time to research which states offer the best lands for homesteading. Some properties may look perfect during the summer but could be untamed during the other seasons.

Know your options and how much land you would need for your homesteading dream. There are country and farm estates to choose from, but do not assume that the zoning will automatically permit farm animals.

It is best to check first. Believe it or not, some rules allow horses but ban rabbits and chickens in certain estates. The closer your land is to the city, the likelier that you are going to have to deal with bylaws on livestock ownership.

Ask the locals about the property that you are considering. Check for running water sources and ample farming space.

Buying Animals Before Building the Fence and Shelter

You are going to care about animals, so they are going to be your responsibility. Make sure that they have ample food, water, and shelter as well as protection from probable predators (e.g., bears, foxes, and wolves).

Many hobby farmers find small animals cute, so they bring them home as if every single one is going to be a pet - do not make this mistake.

Build a proper shelter and fence on your homestead and then add the goat shed, the rabbit hutch, and set up the rest of the barn.

Remember also to build the structures right the first time, or you will be fishing deep into your pocket for restructuring projects.

Zero Breeding Knowledge

Are you going to sell the baby animals that the flock will produce?

You need to plan if you want to include breeding as a part of your homesteading. After all, it is not enough to have the courage to butcher your own animal or to possess the number of the local slaughterhouse. You also need to think about your potential customers and how to deal with the offspring once it is born.

Take note also of the animal feed costs, as these are recurring costs that you need to include in your budget. Furthermore, factor in the cost of gas, equipment, and your time even when you will make your own hay.

These beginner homesteading mistakes can be prevented. There is no need to agonize the way others before you have suffered. If you and your family are just starting out, then this book will help you start your homesteading journey more smoothly.

PART TWO: LIVING OFF THE LAND

Finding Land for Your Homestead

The majority of Americans believe that acres of crops and jam-packed feedlots are necessary to feed everybody. To many, large-scale agriculture with its GMOs and chemicals is needful just so that everyone will not starve to death. Even those who already know that doing agriculture the organic way produces the same results, believe it or not, are swayed to thinking otherwise.

But how much land is needed to feed the nation? How about a family?

As millions of farmers and owners of land would attest, this depends on the land quality as well as the size of the family that you need to support. You can grow food – everything that you need, in fact, on just a small piece of land.

You need to design a system that will work with natural processes to maximize food production. Developing the land for food production can start with as little as two acres (or even less).

Some principles and techniques can even grow food in as little as 25 percent of an acre. If people in the suburbs can live off their land, well, so can you.

Homesteading, as well as self-sufficiency, has become quite popular in recent years. However, the trend should in no way become your sole reason for joining the hundreds who have taken the plunge.

This book will guide you on how to jump-start your homesteading venture by learning intensive gardening strategies, even with less land. It will inspire you to become food-secure as soon as possible. It will prepare you for emergencies that are all realities of homesteading life, but before you set out to acquire the necessary skills and get your boots all muddied, how about buying a piece of land for your homestead first?

So how do you find the land that is perfect for your homesteading dream?

Browsing several real estate listings could get taxing even when you are merely looking for a traditional house – searching for a survival homestead is a different task altogether. Only a few real estate agents have the skills to answer your homesteading needs.

Before you make an appointment to look at a property, first, speak with your bank to ascertain that you are, indeed, ready for a loan. Ask how much your financial institution is willing to lend. Obtain a pre-approval letter before you even make price negotiations with the seller.

There are many options apart from a traditional home, so seek out the agents, preferably, those who are into homestead specific brokerages. If you are having difficulty finding a survival real estate agent in your locality, then find one who already has extensive experience in listing and then selling agricultural properties.

A lovely piece of farmland does not always translate into a survival homestead, so be sure to inspect your options more closely. Bucolic properties will surely be presented to you, and there are certain aspects of this kind of land that you need to scrutinize.

Water Source

If the land – no matter how idyllic it may seem – does not even have a single natural water source, then have it removed from your potential homesteads. Look for a natural pond, spring, lake, or a creek or river frontage because these are the ideal and sustainable sources of homestead water.

If your potential homesteading land answers all your requirements except the water source, then have it shortlisted if its terrain could support one (or more) human-made ponds.

Check the age of the underground wells within the property and do this before making any offers to the seller. If there are no wells, then that is going to be an extra $15,000 to drill one.

If you plan on going off the grid, then you have to be able to fetch water, so be mindful of your water sources manually.

Acreage

Do not be tempted by bargain prices. Take, for instance, fixer-upper farmland that has the potential to become a survival getaway. It could have all the acres that you need (plus some), but if half of it is wooded, and what you need is pastureland, then you have a lot of work to do to make the land adapt to your needs.

It is possible to grow and harvest ample food that can feed a family of four – many accomplished homesteaders have already done it – on a three-acre land (or even less). What you need are the skills to take control of the natural resources and put every single inch of the land to good use.

The ideal land size for a survival homestead is largely dependent on your family size, budget, skills, work availability, livestock types that are being raised, and natural attributes that are needed to sustain a farming lifestyle.

Building your food stock requires at least five acres of homestead land. Here, you will grow livestock such as rabbits, goats, and chickens. You can choose to add turkeys, ducks, pigs, and a cow.

Gardening Plan

Yes, you will need to get used to planning. Plan the size of the garden that can feed your tribe. You need a more extensive design if you want to have a month of the stockpile, and now, with a visual representation of the potential land right before your eyes, you will be better able to draft gardening strategies.

Consider also if you will grow a traditional garden or a vertical one. Will it be necessary to build a greenhouse? Plan where you will situate the garden. As much as possible, make it completely hidden from the road and your neighbors.

Imagine having a long-term disaster, and with your garden in plain view, it could easily become the target for marauders or your neighbors with a starving child.

Get a soil sample also before you sign the contract. Check if the topsoil is sandy, clay, or loam. Make sure that it can sustain the vibrant greens that you see during the inspection.

A Refuge

At the end of the day, the land that you are going to purchase is where you will live, so be able to define what you want. Find land near your home. If you are attached to your local culture, then, by all means, stay there. If your happiness is defined by a sense of community, then find a place with a culture and language that are familiar to you and your family.

Brett R. McLeod, Ph.D. from Antioch University in England, believes in the power of connectivity and localized lifestyle as an alternative to the popular consumer culture.

Visualize the habitat that you will create. Would you want to pick apples and oranges? Would you want to milk your cows? Would you preserve foods in your pantry? The natural, innate tendencies of land may be difficult to assess at first glance, but research will greatly help. Remember that when humans fight nature, more often than not, the latter always wins.

Look for overgrown and weedy fields, neglected lots, choppy terrains, and out-of-the-way spots. The general climate in the area also matters. If you love the sub-tropical heat of California or Florida, then that is the best place for you to set up your homestead.

Visit the courthouse and search for maps. Take time to familiarize yourself with the region's watersheds and topography. Would you prefer flat land or one with rolling hills?

Talk to the locals, such as the farmers, vendors, the feed store owners, delivery personnel, and potential neighbors.

All in all, do not rely on real estate ads – only believe what you see and hear. Ask around and shop around. Inspect roads and look for areas with people who do a bit of farming. Keep in mind that large, cleared lands that can accommodate giant tractors are going to be pricey, and those that are already used in farming might have soils that are saturated with chemicals.

Sometimes the answer to your homesteading requirements sits, neglected, in an abandoned space, so keep an open mind.

Free Land

The depopulation in rural communities, brought on by the influx of people seeking jobs in cities, is now slowly being reversed by the homesteading trend. Free land is now being offered to prospective residents in the following states:

Alaska

Land here is being sold for private ownership and settlement. There is also a program where applicants are allowed to stake a piece of land in a remote area set aside, specifically, for recreation use. These parcels of land are leased for a limited time, and the applicants have a chance to buy them at fair market value after an appraisal and survey.

There is no need to fulfill prove-up or building requirements – you can apply for this program so long as you are a resident of Alaska.

Iowa

Free land is also offered to families who can build a residence in specified lots. The Manilla Economic Development Corporation makes the offer at zero cost for families and individuals.

Marne, IA, also offers free lands averaging 80' x 120'. You may apply via the town's website.

Kansas

There are a handful of communities in this state that are offering free land to draw potential residents. Most of this land is serviced municipal lots that are just right for gardening but may not be best for full-scale farming.

There are also restrictions on house size as well as the time that you need to finish building. Find opportunities in Lincoln, Mankato, Marquette, Tescott, and Osborne.

Minnesota

New Richland offers 86' x 130' land to anyone interested in constructing a new home in just one year. The town has a free land website that you can visit.

Nebraska

The city of Beatrice gives away city land to any individual or family who is willing to construct a home and then live in that home for three years (Homestead Act of 2010).

So far, the program has proven to be successful. Other towns are also offering free land as well as other incentives for their new residents. Find opportunities in Curtis, Central City, Giltner, Elwood, Kenesaw, Juniata, and Loup City. Get in touch with these towns' municipal offices to consider your options.

Other Opportunities

These could also present as rural homes with owners who have to move to assisted living communities. These owners may be looking for future occupants or for people who would want to care for the property at the moment. Think of caretaking as your foot in the door.

Look also for cooperative opportunities. Some large-scale homesteaders might not even be able to afford their lands, were it not for their shared arrangements with other homesteading families.

Cooperative living is not a new arrangement, although it may not be as common as city and town housing co-ops.

Shared garden and orchard spaces are practical means to grow your food initially. You can choose to opt-out later on or be like some of the homesteaders who have stayed in co-op lands for decades.

Lastly, before you put up cash, you also need to check the following:

- Property access rights (make sure that you have legal, permanent transferrable access specified, to the letter, in the deed).

- Sewer and drainage matters.

- Mineral rights (you would want to stipulate what will serve as your compensation should coal or any kind of mineral be discovered on your property).

- Timber rights (check how much timber is granted to you until the land is already paid for).

- Land easements (of course, you would not want to build a vegetable garden right in the middle of a neighbor's right-of-way).

- Permits and zoning.

Protect yourself in all aspects – this is the basic rule to search for that perfect piece of land. Have every transaction recorded at the public registry – as the saying goes, *better safe than sorry.*

Growing Your Own Food

Growing fruits and vegetables need not be overwhelming. It also offers many benefits, whether you are just tending a backyard garden or have committed to a life of farming.

Eating fresh is one of the leading reasons to grow and harvest your own food. You can help improve your family's health as you grow food in your backyard. The vitamin content of these foods will be at their highest levels because they come straight from your garden.

Another reason to grow fruits and vegetables is so you can save some money on groceries. One seed packet costs less than a dollar, but it has the power to shrink your grocery bill later on.

Gardening can also help the planet in so many ways, especially when you decide to say "no" to herbicides and pesticides. Through you – no matter how small your initial contribution would be –, there would be less water and air pollution.

All the weeding, planting, watering, and eventual harvesting are physical activities that can also serve as your outdoor exercise, so go right ahead and start gardening.

The Garden Soil

Soil is a crucial factor for every successful garden, regardless of the kind of garden that you plan to grow. It will take time and research to find the right kind of soil, test it, and then finally use it to grow your plants.

There are two extreme soil textures – sandy and clay soils. Coarser sand particles are known as gravel. Touch it, and it will easily crumble. Clay soil, on the other hand, is smoother. When wet, clay is sticky and will easily smear like paint.

When clay and sand are mixed, this is the ideal soil for gardening, which is known as loam. Sandy loam permits the drainage of surplus water while clay loam holds water. Remember that the ideal gardening soil has equal parts of water and air.

When you hear about organic matter, this is simply the vegetable and animal matter that is undergoing the decomposition process. Adding humus offers little amounts of chemicals that are needed for plant growth. There is also ample food for the bacteria in the soil, which, in turn, provides organic chemicals to the growing plants.

Humus is what you need to keep the garden soil porous and open. This allows water and air to penetrate the soil, while also acting as a sponge that keeps the roots healthy so they can hold nutrients.

What Else Do Your Plants Need?

Plants need carbon, oxygen, and hydrogen in order to grow. They get all these from water and air together with the soil. They also need phosphorus, nitrogen, and potassium that can be supplied by fertilizers.

Other needful elements are magnesium, sulfur, iron, and calcium. These are readily available in most soils. Trace elements like zinc, manganese, boron, copper, and molybdenum can be added in minute quantities.

All these elements must be present, as insoluble forms, in the soil as well as the water and air.

There are three elements that every fertilizer must have so that it can be classified as complete: nitrogen, potassium, and phosphorus. Take note that, by law, the amounts that the fertilizer contains of each element must be clearly indicated on the package. If boron is added, then the number will also be indicated on the packaging.

Nitrogen is derived from slaughterhouse by-products like bone and blood, phosphorus is processed from phosphate rock, and potassium comes from wood ashes.

Another thing that you also need to check concerning the soil is whether it is acidic or alkaline. Garden soil can be either or neutral. Most vegetable and fruit plants thrive in neutral or moderately acidic soil. There are even plants that demand acidic soil, while fewer plants demand sweet or alkaline soil.

You also need to check the subsoil that is the layer right beneath the surface. It can be just a few inches down, or it could go as deep as 20 inches. This can be full of gravel or sand; that is why you need to check how deep this layer is as plants only get the nutrients from the topsoil.

Sandy or clay subsoil is difficult to work with because it requires a drainage system. Sandy subsoil can, however, be corrected by adding peat moss, manure, and organic matter. Planting green manure crops such as oats, rye, ryegrass, clover, and vetch can also improve sandy subsoil.

Other ways to improve the soil is to mix sand with clay soil and vice versa. An application of mixed cow and horse manure (about two-three inches thick) is the fastest method to improve the soil and add plant foods. The manure has to be partially rotten already.

Use sheep and chicken manure only when mixed with compost or dried leaves. If manure is hard to get, green manure and peat moss are good alternatives.

The Compost Pile

The foundation of a compost pile is comprised of the following:

1. Dead stems and stalks of corn, marigold, and zinnia stay at the bottom.

2. The next layer is six-inch deep grass cutting made with dead leaves, rotting peels and fruits, and vegetable wastes.

3. Next, a three-inch deep garden soil.

4. The fourth layer is a 3-6" deep peat moss.

5. Lime

6. Manure

7. The top layer is the complete fertilizer.

Starting at the second layer, repeat the sequence until the heap has reached four-six feet. The sides of the heap should be sloping inward, with the top left part depressed to hold water.

Upon completion, the pile should be watered. Make sure that every layer is saturated with water, but not to the point of overflowing.

Things you can add to your compost pile:

- Dog, fish, or cat food
- Fur from brushing the dogs
- Alfalfa hay
- Bee droppings
- Cow, horse, or goat manure
- Aquarium plants
- Bird droppings
- Fish scraps
- Horsehair
- Feathers

Apart from the pet stuff, you can also add laundry room things such as:

- Dryer lint
- Shredded denim or cotton clothing
- Shredded old towels

- Shredded old socks

Kitchen items that can also be thrown onto the pile:

- Paper napkins
- Paper towels
- Vegetables that have been freezer burned
- Teabags
- Cooked rice and pasta
- Coffee grounds
- Coffee filters
- Stale cereal
- Stale chips, bread, and crackers
- Chopped wine corks
- Nutshells, except walnuts
- Shredded cardboard and food boxes
- Popcorn
- Stale beer
- Old spices
- Old wine
- Eggshells
- Outdated yogurt
- Corn cobs
- Paper receipts
- Leftover salads
- Chopped up seeds
- Pizza crusts
- Any paper wrapper

Even office trash can be added to the compost pile:

- Shredded papers and bills
- Sticky notes
- Pencil shavings
- Paper envelopes (now the window type ones)
- Matte business cards

And, of course, household items that you can also turn into compost:

- Newspapers
- Used potpourri
- Contents of a vacuum cleaner
- Houseplant leaves
- Dried flowers from vases
- Fireplace ashes
- Used tissue paper
- Shredded toilet paper and their cardboard tubes
- Old leather gloves
- Shredded old sheets
- Glue
- Razor trimming
- Leather wallets and watch bands
- Scraps of Ivory soap
- Cotton balls
- Greeting cards
- Used matches
- Fingernail and toenail clippings

Fruits and Vegetables

There is nothing like enjoying a fresh bowl of hand-tossed salad with the ingredients picked from your garden. To begin enjoying these freshly picked vegetables, you need to find the right location that gets ample sunlight during the day but also has shade during the hottest parts of the afternoon. The shading could be any permanent structure or a towering tree. You can also use shade cloth if you do not have either.

For small spaces, you can grow your vegetable garden vertically. The best plants that benefit from vertical growth are: beans, peas, cucumbers, vining flowers, small squash, and tomatoes.

You will need a trellis (wooden or metal bars used to support climbing plants) for these vegetables so you can better prune and harvest them.

Growing vegetables in pots is also a thing in many apartments and tiny homes. You can begin with these for now, and then you could eventually look for an empty neighborhood lot that you can convert to a community garden.

Even newbies can start planting fast-growing vegetables soon. The shorter the time between the planting and harvest times, the more fast-growing the vegetable variety is.

Other factors that influence a vegetable's growth are sunlight, soil nutrients, ample water, and the time and energy that you put into caring for them. Now, would you like to learn the easiest vegetables to plant in your new garden?

Zucchini

This is, perhaps, the most fail-proof vegetable to grow on your homestead. This is a low-maintenance plant that grows in almost any kind of soil and climate. It's the perfect SHTF crop.

Do not let it grow too big, though. Freeze any excess crops for later use in baking bread and preparing soups and sauces.

Zucchini can also be grown on rooftop gardens (great for apartment homesteaders) and in pots.

Greens

Remember the leafy greens like lettuce, kale, arugula, and spinach because they are awesome for beginner homesteaders. These mature in as little as three-four weeks; to harvest, just snip the tops or pick some leaves.

Spinach is a delicious crop that grows best in cool weather, so it is best to plant the initial crop four weeks before the final frost date and then the second crop about six-eight weeks after the final frost date. You can harvest spinach in as early as 20 days. Just pick the outer leaves so you can continuously harvest into the season.

Kale, on the other hand, is a nutritious, sweet-tasting green that is also best planted during the cooler weather. It can even grow during winter. A spring crop can be planted four weeks before the last frost date, with the second crop planted in late summer or the start of fall. You can harvest kale in 60 days when it has fully matured. Young leaves can also be harvested before this, but be sure to leave four or more leaves so that it can still grow.

Arugula has a unique taste that is best used in salad and sometimes with chicken and other meat varieties. Plant the arugula just a few weeks before the last frost date and then the second crop during fall. As soon as you see flowers, remove them so that the plant can grow further. It only takes 20 days for the young leaves to develop – 40 days if you want a mature harvest.

Lettuce is a salad staple, but it is also an awesome leafy green that can be included in any kind of diet. Just like its cousins, it grows best in cold weather, so plant the spring crops about four weeks before the last frost date. The fall crops can be planted at about four weeks before the initial frost date.

Lettuce takes 40 days to mature, although you can harvest leaves even when they are still small.

Beans

These vegetables are nutritionally charged foods that can be eaten in different ways. They are easily dried, edible even when raw, and they are a staple in any vegetable garden.

Plant beans when you're sure that the soil is going to be consistently warm. Many beans mature in just 60 days, although some can take as long as a hundred days. Farmers prefer to use a trellis when growing beans for greater support and tastier fruits.

Snap Peas

This is a prolific producer. Peas come in different varieties, but the snap peas are great for starter homesteads because they can grow all-year-round. Plant them one-two weeks before the last frost date. Harvest within 40-60 days.

Green Onions

These crops do not just add flavor to your meals; they are also highly nutritious. Just like the snap peas, green onion prefers warmer soil, so make sure that the last frost date has passed before you plant it.

It only takes 50 days to mature so you can uproot it then; although, you can harvest the stalks in between. Limit your stalk harvests as these can restrict the bulb's growth.

Carrots

These root crops need loose, sandy soil, so take it easy using manure type fertilizers when you plant them. It is best to plant lots of carrots as growing them can be a tad tricky. They are a great practice crop for you, though, since their growth happens away from your site.

Plant carrots four weeks before the last frost date and then harvest 60 days after

Turnips

An excellent source of antioxidants and fiber, this root crop is a tough and reliable garden staple. Some use it as a substitute for potatoes.

Plant it three weeks before the last frost date then harvest in spring. You can plant again in September for your fall harvest. It only takes 30-60 days to mature, with turnip greens harvested in 30 days and the mature crops in 60 days.

Tomatoes

But what's a homestead without a patch of tomatoes?

Wait for 60 days before you can harvest. This is after planting after the last frost date in spring and then pruning every seven-ten days.

Corn

If you have bigger land, then you can have stocks of cornmeal ground from your own corn plants. The sweet variety can be eaten right away, canned, or even frozen.

Since this is your first time, buy seeds from the market and then plant with a 20 to 30-inch row space. Plant about 33,000-38,000 seeds in every acre. Increase the rate, and you can maximize the corn yield.

Always check the weather report. Plant a small amount during the dry season and more during the wetter months. Scientifically, the temperature should be no less than 50 degrees Fahrenheit. Make sure that there are going to be five to 14 days of warm days ahead. The soil's moisture must not be wet, or it will be too compacted, and the corn will not grow.

Add organic fertilizers according to your plants' needs. If the soil is fertile, then there is no need to do so.

Medicinal Herbs

Herbs are useful plants that can boost everyone's health and wellness. Most of these are not difficult to plant; there are many with lovely flowers and unique foliage. They can also be easily infused in any garden bed; in fact, some can grow pretty well in pots.

The basic arsenal would be garlic, basil, and mint – start small, so it will be hard to miss. All these are also used in cooking, so you are fulfilling both gastronomic and wellness needs.

You can start with:

- Ginger
- Licorice root
- Ginseng
- Myrrh
- Fennel
- Walnut

And what about fruits?

Berries

Blackberries, blueberries, and raspberries are all fruits that are rich in antioxidants. These are considered superfoods because of their nutritional values.

Black cherries are great liquor and jam ingredients. The tree grows quickly at two-three years, so you definitely have something to look forward to.

It can take one-three years for the bushes to produce their first fruits, but you can harvest annually after that. So, without a doubt, berries are a good investment for year-round consumption.

Other Fast-Growing Fruit Trees

Fig trees are known to bear fruit after only one-two years of planting. They are easy to grow and can even be planted in pots when you live in colder regions.

Mulberries are known to produce fruits in just one year. You can also share this with the ducks, chickens, and turkeys.

Peaches can produce fruits in less than three years. What they require is well-drained soil and a partner so they can cross-pollinate.

Citrus trees can be grown in pots should you have limited gardening space. Choose from oranges, lemons, and limes. These are self-pollinating, so there is no need to worry about buying a pair to produce fruits.

These are the basic vegetables and fruits that you can plant with your starter garden. Remember that you can begin gardening even with just a small parcel of land.

Live in an apartment? Then find a bucket and start planting. There is simply no excuse – the best time to enjoy the blessings of homesteading is now.

21 Homestead Gardening Hacks

Now that you learned which vegetables and fruits are the easiest to plant and harvest, you are sure to reap the blessings of your hard work soon. In the meantime, here are some homesteading hacks that will further increase your harvest.

#1. Use Wood Pallets

These old pallets can be turned into vertical gardens or raised beds. Those who have smaller acres of land or are just homesteading in their backyards would do well to grow peas, peppers, lettuce, basil, mint, and even cucumbers using old wooden pallets.

To make a raised garden bed, you will need:

- Wooden pallets
- Topsoil
- Scissors
- Garden fabrics
- Staple gun with staples
- Seeds and plants

Method:

1. Clean the pallets and then turn them over.

2. Lay the fabric across the back of the wooden pallet.

3. Staple the sides, making sure that the fabric is taut.

4. Turn the pallet over and check for any holes where dirt could leak.

5. Fill the pallet beds with topsoil from your garden or a mixture of prime soils that you bought.

6. Start planting as you would in a regular garden.

#2. Hang Shiny Pots, CDs, and DVDs

Believe it or not, keeping squirrels and birds away from your fruit trees is as easy as gathering some old CDs and DVDs and then hanging them right on the tree. These dangling discs will reflect light and act as prisms so that the birds will think that a larger animal is already present.

Squirrels do not like the reflecting light, so they stay away from such trees.

#3. Pinch off the Top

While it is awfully tempting to grow your herbs and vegetables to full maturity, it pays if you pinch off the top stems now and then. This simple action encourages further growth because the dormant leaves are signaled to grow outwards.

#4. Use Kitchen Scraps

If you still have not started a compost heap, then put those kitchen scraps to good use. Throw the scraps (minus citrus fruits and fatty meats) into the blender, add a bit of water, and then mix away. This mixture will easily decompose so you can then put it at the base of plants.

#5. Add a Drop of Oil

You can prevent mosquitoes and other insects from breeding when you add just a few vegetable oil drops to the water. This will form a thin layer over the water surface that will deter the insects but will not harm the birds.

Use this method on your water barrels and birdbaths.

#6. Dwarf the Fruit Trees

By now, you already know that you do not need huge amounts of land to grow vegetables and fruits. There are now dwarf fruit trees that you can plant even inside your home.

Growing these fruit tree varieties will even allow you to take care of non-native fruits. Choose from dwarf orange, apple, coffee, banana, and lemon.

#7. Fertilize with Poultry

Duck and chicken manure make good fertilizer. As you prep the plots for your garden, you could set up a temporary pen for the chickens. They can be kept in there for a week (up to two) depending on the plot size. The chickens will naturally fertilize the soil while they also peck at pests.

You can also collect rabbit droppings by creating a tub filled with dirt. Place this underneath the hutches. What you collect will become a nutrient-rich soil that can be used as seed starters.

#8. Create Rain Barrel Systems

Homesteaders prefer to have rainwater harvested because tap water potentially has high levels of chlorine. You can even save a huge amount of money since you can collect as much as 300 gallons of rainwater for every inch of rain on a 500-square-foot roof.

To make your rain barrel system, prepare these materials:

- A large plastic garbage bin (or more)
- A tube of sealant or Teflon tape
- Two rubber washers
- One hose clamp
- One spigot
- Drill
- Landscaping fabric

Method:

1. Drill a hole in the garbage bin. This is where the spigot will be inserted. Be sure to use a drill bit that is smaller than the spigot's size. Do not place the hole too low into the bin so that there will still be ample space underneath to fill the barrel.

2. Insert the spigot after placing a metal washer into its threaded end. Fit with a rubber washer on the threads so that the washer is held in place. Check for holes that could potentially leak.

3. Seal the rubber washer with a waterproof sealant. Make sure that the spigot is secured with a hose clamp so that it won't come loose from the rainwater barrel.

4. Create a hole that will gather the rainwater from the downspout. Drill another hole at the top where the rainwater can overflow.

5. If you do not want to waste overflow water, then you can connect a PVC pipe or hose to another barrel. This way, any excess water will run to the following barrel.

6. Cut the landscaping fabric and place it over the top of the barrel. This will serve as a barrier against mosquitoes and other insects that could get into your rainwater barrel.

7. Put the lid on, then close it.

8. Place the rain barrel underneath the downspout then wait for the rains to fall.

#9. Use Toilet Paper Tubes as Seed Starters

These tubes are biodegradable, and they come at the right size for seeds to grow in. To use, cut flaps on one end of the tube then close it off. This can then be used as a home for your starting seeds.

#10. Reuse Coffee Grounds

Old coffee grounds can be used to enrich the garden soil while it also deters aphids from damaging your leafy greens.

#11. Make Beer Traps

Beer can be used as a trap for garden snails and slugs. Have a shallow dish filled with beer then put these in your garden. The scent of the fermented liquid will lure the slugs into the dish. While they can climb into the drink, they will not be able to ascend out of the trap.

#12. Plant Densely

Apart from aphids and slugs, one other thing that you wouldn't want growing in your garden is weeds. To keep them from appearing in the first place, you can plant densely on the garden beds. Mother Nature does not like gaps, so she quickly fills any empty spot. Plant densely, and there would be no room for the weeds to grow in.

#13. Find Uses for Wood Ash

As you drive away the garden pests, you can also use wood ash to fertilize your growing vegetables and fruits. This is packed with potassium, calcium, magnesium, phosphorus, sodium, iron, and zinc. To use, dust it over the plant bases or place it together with other compost ingredients.

Creating a chicken run using an old tire filled with wood ash will also keep the hens from being bored while keeping mite infestation to a bare minimum.

#14. Grow Roots with Honey

Honey can also be used as a root grower. Instead of purchasing rooting hormones, you can apply honey on the plant cuttings. Its antibacterial properties will help in propagating the roots more quickly.

#15. Use Packing Peanuts in Large Pots

Instead of filling up large pots with pure potting soil, try mixing it with packing peanuts. You will use less soil while improving the water drainage. You will also find that the pot has become lighter the moment you decide to move it.

#16. Never Dig Wet Soil

Soil is not just difficult to dig when it is wet. Digging wet will also guarantee that it will be compacted (this is not good for gardening). If the soil sticks to your boots' soles when you go out and dig, then postpone the task to a drier day.

#17. Keep Rust at Bay

To protect your gardening tools from rust, use a bucket or a big terracotta pot. Fill it with abrasive sand with some mineral oil. Place your tools into the bucket once you're done gardening. The abrasion and oil will keep them sharp, clean, and lubricated.

#18. Soak the Seeds Before Planting

This is an effective way to germinate seeds quickly. Soak them overnight before planting them. Do not soak for more than 12 hours, though, because this will result in decomposition rather than faster germination.

#19. Diapers for Water Retention

If you live in a region with a hot climate, then this hack is especially useful to you. Place a clean diaper (with the absorbent side up) at the bottom of a garden pot. Fill the pot with soil, put in your plant, and then water.

#20. Apple Cider Vinegar as a Supplement

One of the best ways to boost your livestock's immune system is to add some apple cider vinegar to their food or water during winter. You soon see that the animal/s will be less sickly or will not suffer from infections as you boost their immune system naturally.

#21. Rotate the Crops

Crop rotation is the process of growing certain vegetables in specific plots every year. An example is when you plant brassicas right after planting legumes the previous planting season. Since the legumes will have already added nitrogen to the ground, then the brassicas will greatly benefit from the nutrient.

Crop rotation avoids the depletion of soil nutrients.

These are just 21 mind-blowing gardening hacks that you can start practicing in your quest to getting more involved in the cycle of life. So, are you now ready to learn about livestock farming?

Self-Sufficient Farming: What Livestock and How Many?

Throughout the centuries, farmers all over the globe have developed various livestock breeds. One breed (or even more) becomes extinct with each passing month, with 20 percent of the world's population of pigs, cattle, goats, horses, and poultry breeds now at risk of extinction.

If you went back in history, you would see that farmers used to operate in smaller acreages. Though this was the case, they still took care of a wide variety of animal and plant species. And the breeds that were developed often answered specific needs.

The Texas Longhorn is an example of a breed of cattle that was developed from Spanish cattle. These arrived during the earliest colonization days of North America. The ranchers then needed tough cattle that could live through predator pressure, poor forage, and still be able to produce healthy calves.

The unique horns soon emerged and were useful in fending off predators.

There is no need to create a unique breed in your homestead now. What you need to learn is to look for the animal breeds that will be productive and sustainable. The livestock farming tips that you will learn here are general guidelines only. If you live in an arid region, then Ossabaw Island pigs are perfect animals to grow on your homestead. Looking for chickens that can lay eggs even during the harshest winters? Then raise Chanticleer chickens as livestock.

Animals You Can Raise on Your Large Homestead

Rabbits

Rabbit meat is tasty and has low cholesterol. The animal can be raised even in a small backyard, and unlike chickens that need special equipment to be raised, the rabbit is not only cheaper to feed, but it also does not require a huge space. It should be confined, though, so that no predators can get near it.

Rabbit hutches must be built above the ground so that skunks and dogs are kept at bay.

If you want to harvest more than the rabbit meat, then you can also use the fur. The rabbit pelt has been worn for years. Angora rabbits can shed enough fur to make a coat.

Make sure that the hutches can effectively protect the rabbits from snow, rain, and predators, such as dogs and skunks. For warmth, you can add straw or some bedding materials to the hutches.

Note that a full-time rabbit business should have 600 does and around 60 bucks. Every doe gives birth to 25-50 live rabbits every year. To feed a family of five who wants to consume rabbit three times each week, you should have 12-13 fryers (baby rabbits) per month.

Goats

Modern homesteaders can attest to the level of self-sufficiency that this livestock can provide. You can now have stocks of cheese and milk provided by your dairy goats.

Note that the goats need to give birth to kids so that they will lactate. You can then milk them throughout the year (for up to three years). You can plan and stagger the kidding so that you will have milk throughout the year.

One average-size dairy goat can provide three-four quarts of milk every day. Depending on the goat milk's butterfat content, you can even get up to one pound of cheese for every gallon of milk.

You can also raise goats for their fur. Called fiber goats, they can provide yarns that can be spun as blankets, hats, sweaters, and other items. Spinners pay for goat fibers, so while you put the weed-killing abilities of this livestock to good use, you even earn money from the fibers that you sell.

Build durable fences when you want to raise goats. Use woven wire, cattle panels, or welded wire. The fence should be no less than four feet high. Tighten the materials so that the goats cannot destroy them. A strand of electric wire can be strung across the top of the fence.

Every goat needs 200 square feet of fenced run. This is the minimum, so you can make room for your goats to roam and play. They also need a constant supply of fresh water.

Does, bucks/billys, and pregnant goats need about three cups of feed every day. When being milked, allow the goat to eat more than this. Wethers eat less than a cup.

Goats prefer weed, hay, and alfalfa. Forage should be a part of their diet, including leaves, dandelions, and weeds.

Make sure your goats have ample vitamins and minerals, too. Their vitamins are not the same as sheep, so check the labels. They also need routine care, such as worming, hoof trimming, and vaccinations (especially for tetanus).

One to two goats can easily provide an annual supply of fresh milk for a family of five. To grow your goat livestock, you can start with two does and one ram. Double this population if you have a bigger family to sustain.

Sheep

Sheep are great for wool harvests, meat provision, and milk production. If you have goats, then you have, technically, enough training to raise sheep. They are alike in terms of their shelter and fence requirements.

Salt blocks are great sources of minerals for sheep. Fresh water is also a must, so if you can, set up a water station for the sheep.

Sheep cannot consume copper – this is poisonous to them.

General care for them includes trimming their hooves, shaving their wool, worming, and checking for other parasites.

It is suggested that a large family would begin with two ewes and one ram. This can eventually be grown to four ewes.

Pigs

These are a great addition to any homestead because they eat up the often problematic kitchen leftovers, plus they provide delicious pork. Just like goats, they constantly break fences in their need to escape.

They are easier to farm compared to cattle, though, because they require the lowest maintenance. They do not need immunizations, they don't have to be milked, there's no need to sheer them, and you don't even have to be that hands-on.

Provide them with ample feed and water, and you're good to go. Make sure that they are also fenced in together with their big buddies.

A pair of pigs would be enough for families with six-eight members.

Cows

Here is something to wow you – a thousand-pound cattle averages 430 pounds of steaks, ground beef, roasts, stew beef, etc. This should feed about 430 people who would consume one-pound serving each.

You can raise cows for meat and milk. Calves can also be sold to homesteaders like you. Eventually, even their hide can be turned into something useful. For many farmers and homesteaders, this is dream livestock. Prepare yourself because these are bigger animals, so they require a larger space and infrastructure that can contain them.

Getting a cow requires an upfront cost. You also need to prepare for the maintenance that it will need.

Now the question to ask yourself is: "Why will I raise a cow?"

You need to know your goals. Raising the cows solely for milk will be costly. First, they begin producing milk when they are two years and older or when they have their first calf. After the colostrum (first milk) is given to the calf, the cow can then be milked for two years.

Raw milk is rarely sold in retail stores – only a few states allow its sales. If you live in one of the 20 states that consider unpasteurized milk sales to be illegal (e.g., Alabama, Virginia, Colorado, etc.), then it will not be feasible to earn much money from milking.

A single cow can provide you more milk than what you can normally consume. But if you are only buying two gallons of milk at $3 apiece, then you're not saving a lot there.

You could make more money if you raise the cow and eventually sell it.

For homesteaders who purchase cows for their homegrown meat, it is necessary to find a meat processor or custom slaughterers. Grass-fed cows are ready to be slaughtered in 28-30 months while the grain-fed breed is ready in 15-16 months.

Invest in stock or a heifer that weighs 600-700 pounds. It will grow during the summer, and then it will be ready to be stocked in your freezer by fall. It can be challenging to process the meat since qualified processors are few. Schedule a slot for the slaughter date way before it happens.

If you plan on selling beef, then you also have to research your potential markets. You have to label and safely process the meat before you can sell it. Remember that meat is generally more lucrative than milk.

Different breeds of cows thrive in different climates. The hot climate of Florida or Texas will make the Brahman cow thrive. Red Angus will live in North Carolina.

Raise the Angus breed, Belted Galloway, or Hereford (all small breeds) if you live in an area where feed is pricey. Always buy cows from reputable breeders and feed from trustworthy suppliers.

One cow needs about an acre of land. Once they begin having calves, then you need two acres for each cow-calf duo. Check the forage quality on your land. You might need to add supplements to their diet if the forage quality is poor.

Feed the cows with grass and legumes and then balance their diet with mineral supplements. Vetch is also good in winter. If you want to add grains, then use high-protein corns.

Just like goats, they also need a constant supply of fresh water.

Fowls

Ducks' space requirements depend on the breed that you are going to raise. Larger breeds, such as Saxony or Jumbo Pekin, would do well in a space of four-five square feet. For them to be able to run around, you will need to set up a minimum of ten square feet for every duck.

Pekins are heavy, so they are also highly recommended for homesteaders who worry that they will fly away during winter. These are also wonderful foragers, so they grow up to a size that will provide ample meat for the family.

You can also create a small pond so they can splash around. If this is not feasible at the moment, then allow a few ducks to share the kiddie pool.

Be sure to keep any fowl livestock protected. Ducks are commonly preyed upon. Even your dog might chase the ducks, so prepare for this scenario. Place a fence all around the area where the ducks are supposed to roam.

Hawks will have a hard time going into the fenced area since they need to swoop down instead of lower themselves like a helicopter.

Wooden pallets can be set up as duck houses. They are perfect for airflow and moisture. Build them low to the ground so that the ducks will be enticed to use them.

Ducks rise earlier than chickens, are hardier, and lay bigger eggs. They do not lay eggs throughout the year, though, so enjoy the fresh duck eggs while they are being provided to you.

Another type of fowl that you can raise on a homestead is the goose. Just like the ducks, you need to keep the geese happy if you want to harvest ample meat. They are easy to care for, they eat less than ducks and chickens, and they feed mostly on grass and then some corn and oats during winter.

You can simply let nature take its course when you purchase a pair of mature adults (two years old or older). They are monogamous animals, so the gander will often pair with just two female geese. Let them live together for a few months so they can begin to mate.

When properly mated, a female can lay 20 eggs in spring. She will only be able to hatch 12 to 15. You can take six-eight of her eggs and help the female hatch them using an incubator. As goslings are hatched, it could take as long as 24 hours for each one to break out of the shell. Never help it.

Goslings and other geese should be transferred to the fenced run as soon as they can mingle. Set up a fence that is no taller than three feet. If you have a dog that constantly preys on them, then raise the fence higher.

The geese can weed out a strawberry patch so you can fence them in there. Keep them out of there once the strawberry fruits begin to ripen. Keep them safe from foxes, owls, and raccoons.

The myth that geese need a large pond or lake to survive on a homestead had been long busted. You don't even need a huge body of water to keep the birds.

Just set up a trough for the goslings and the adult birds to splash in now and then and also for them to drink from occasionally. Geese forage for bugs and grass, so they do not need supplements, except during their first few weeks of life and in winter.

Geese varieties include the Chinese watchdog breed. They are elegant, but they are quite noisy. The Embden variety could weigh as much as 26 pounds, and they produce large eggs.

The Toulouse variety can also weigh as much as the Embden. At breeding age, they can be worth $100-150 each, so when they are sold in pairs, they could give you good amounts of cash.

Quails can also be a part of your large-scale homestead. The common breed – Bobwhite – matures in just 16 weeks, which means you can have quail meat in just four months. They lay eggs at 24 months.

Quails do not require much space. A two feet by two feet by eight feet pen could house as many as 25 quails. Invest in an incubator if you want to profit from quail eggs and meat.

Note that there is a special chapter dedicated to raising backyard chickens.

You will need about two gallons or more of water, every day, for a dozen turkeys. A 4' long waterer can sustain up to 12 birds. A hundred pounds of feed should be good enough for a dozen of them. By the time that the birds mature, though, you will already need to feed them a pound of feed, each, daily.

Bees

Becoming an apiarist (beekeeper) means understanding the biology of these insects. Unfertile female bees are the workers while the queen lays the eggs. There are just a few drones (drones) in the hive. Every worker lives for six weeks, and then she simply dies from working herself to death. Every worker will have gathered nectar that can make a 1/12 teaspoon of honey.

Rev. L.L. Langstroth first conceptualized the modern beehive in 1851. Before this, farmers just left the bees in hollow logs, boxes, or straw keeps. The bees are eventually killed or violently ripped out of the honeycomb. Langstroth made movable frames as well as uniform bee spaces. The interiors have sections that are separated by 5/16". A homesteader can then harvest honey without damaging their home or hurting them.

You can buy established hives from expert beekeepers. This could cost anywhere from $50 to $100. Never buy a colony that has not been inspected by the state department of agriculture or an apiarist (beekeeper).

You can also have bees delivered by mail. The three-pound box can be delivered to your homestead, and that will contain about 10,000 workers and one mated, laying queen.

Visit the bees only on a sunny day. Wear white or any light-colored clothing. Never wear wool. Tuck your blouse or shirt into your pants and the pant legs into your socks.

Do not forget to wear a hat, gloves, and a bee veil. Don't wear perfume and make sure you are not reeking of farm/animal odors. Wash the beekeeping outfits regularly because any residual odor of past stings could trigger an attack alarm.

Use a smoker. Work barehanded only when you are already confident in removing your gloves. Smoke down the frames, then pry and lift a frame. Be careful and gentle with your movements, as any sudden, jerky actions will alarm the bees.

Small-Scale Livestock

From the above list, urban and small-scale homesteaders can raise the following:

- Chickens
- Quails
- Ducks
- Rabbits
- Pygmy goats
- Dexter cattle
- Bees

Chickens are naturally quiet, and they are highly sought for their eggs and meat. Backyard varieties include the Cornish Cross, Freedom Ranger, and Red Ranger.

Ducks are great garden companions – this is if you're juggling tending livestock and growing vegetables. Small breeds are Rouens, Indian Runners, Blue Swedish, and Khaki Campbells.

Quails also have a small variety, the Coturnix. These can lay four-six eggs a week and will sell up to 50 cents per egg.

Champagne d'Argent, California, New Zealand, and Crème d'Argent can be bred and kept in a small area. When you plan accordingly, a small backyard could produce as many as a few hundred rabbits.

Animals for Urban Homesteaders

In an urban setting, the pygmy goat can thrive because it is about half the size of a regular goat. The Nigerian Dwarfs can produce as much as one-two pints of milk. Their meat can still be harvested.

The Dexter cattle can be yours if you have at least half an acre of land. This land must have green grass where the animal could graze. Give it 18-24 months, and its 500-pound meat is ready for consumption. In the meantime, you can milk it for up to three gallons of fresh milk.

These same animals can be raised in community gardens, too. These are the spaces for people who want to homestead even when they live in the middle of the city. There are mobile slaughterhouses that can be commissioned when it is time for the animals to be harvested for meat.

Lastly, urban beekeepers can raise bees on their apartment rooftops.

Just remember to remain on the right side of the law when urban homesteading. If the local government allows livestock to be raised within the city limits, then that is great – just learn the regulations and requirements for all homesteading opportunities.

How to Raise Backyard Chickens

With homesteading as your new passion, you are probably searching for ways to get those feet wet. Buying chickens is a great place to start your journey. These fowls are one of the sturdy foundations of homesteading, and this is because they can fulfill several goals.

What chickens traditionally offered factory farming has blurred. The non-negotiable life of self-sustenance, simplicity, and efficiency is now a life of options and convenience. Chickens can help you regain the more beautiful aspects of homesteading.

Benefits of Raising Backyard Chickens

There are many other things that chickens can offer apart from eggs and meat. Of course, you cannot discredit the healthy option that free-range eggs provide. Store-bought eggs are not just bland; they barely provide the nutrients that your body needs.

Real chickens that roam freely act as chickens do, and they eat what chickens are supposed to eat. Layer feed is a wonderful supplement, but the chickens are meant to feed on proteins and greens.

Eggs produced by organic chickens contain more vitamins A and E. They also have omega-3 fatty acids and have lower fat and cholesterol.

A bird that is allowed to roam freely in its natural habitat also balances all the other elements on your homestead. Chickens can even help get rid of pests and parasites, plus they are also amazing gardeners as they love to nourish themselves with grub and the insects that feed on plants. Just make sure that they do not feed on the vegetables (warning, they like tomatoes).

And what about those mosquitoes that you worry about, as well as ticks and bugs? The risks of heartworm, malaria, and dengue virus are drastically cut down thanks to your feathered friends.

Chickens also naturally aerate the earth by pecking. This leaves a looser soil, so plants are better able to grow. They can aid in fertilizing the soil so you can look forward to greener gardens. Chicken manure can aid in increasing the potassium, nitrogen, and phosphorus contents of the soil.

To best use chicken manure, have it composted for about four to 12 months before you add it to your garden. You can speed up this process by rotating periodically so that the airflow is increased and the compost decomposes quicker.

Another benefit of raising chickens in your backyard is that you can harvest them for their meat. Free-range poultry tastes much different from factory farm chickens – and it does look different, too, with its smaller (yet tastier) breasts, chicken skin colors, and leaner meat. The store-bought chicken looks almost perfect because it was engineered to look that way – now doesn't that scare you?

There is also a sense of pride that comes with being able to provide for your family's needs. As soon as you understand the many benefits of setting the foundation of your homestead by nurturing chickens, then you will have an even higher appreciation for homesteading.

The Meat Bird

Before you buy the chicks, first think of the purpose that they are going to serve. There are laying hens, and there are fighting cocks, but raising fast-growing fowls is a different arena.

With more chickens in your backyard, there is going to be more poop, so it's crucial to ask yourself now if you can handle this scenario. Also, are you okay with saying goodbye to the chickens every six-eight weeks? Would there be a problem with your kids turning them into pets and then not being able to let go when it is time to harvest the chicken meat?

To choose a meat bird variety, you could consider the Cornish Rocks that are a cross between the White Rock and Cornish. These can grow quickly with feed so you can harvest more meat as they increase in size.

You will likely purchase your meat birds as day-old chicks from the feed store or hatchery. Keep in mind that chicks require a lot of care. There should be a brooder area set up for them, with a heat lamp constantly keeping them warm. The brooder temperature also needs to be closely monitored.

Chickens can be kept in a coop (you'll learn about making one in a few minutes) with an attached small run. Have them run free, and you are raising happier chickens, so they are better able to produce omega-3s.

After the birds have grown to their full size (about five-seven pounds depending on whether they are roasters or broilers), it is time to harvest. You can slaughter the chickens right on your homestead or have them transported to the slaughterhouse and then processed.

Some farmers' markets buy birds, but you need to have the birds taken to a USDA-approved slaughterhouse or processing facility. There are mobile facilities in some states, so you can also opt for those.

The Broody Hens and Eggs

If you are looking for laying hens so you can harvest eggs and eventually their meat, then you also need to learn how to pick the best breed, and how to house them and provide the supplies that they need.

You should raise free-range chickens, but if they need to be cooped and just be let out now and then because of the space limitations on your homestead, then that is fine, too.

Urban and suburban settings require fowl confinement, but even these birds need some fresh air and lots of sunshine, so you need to plan their living quarters, too.

Did you know that there are more than 200 chicken varieties nowadays? And you need to consider factors such as the climate, the breed temperament, egg-laying levels, and meat growing capabilities before you should set out to buy one.

If you are aiming for pure breed egg producers, then you can opt for Buff Orpingtons. Your other options include the:

- Leghorn
- Rhode Island Red
- Australorp
- Plymouth Rock
- Sussex
- Wyandotte
- Ameraucana
- Marans
- New Hampshire
- Silkie (this looks cute so be careful about turning them into pets)
- Brahma
- Cochin

- Welsummer
- ISA Brown
- Polish
- Jersey Giant
- Hamburg
- Minorca
- Barnevelder and the list could go on and on.

You can mix breeds if you are aiming for both meat and egg varieties. You will also need a rooster if you want to hatch chicks, but if you only want chickens for their eggs, then you do not need one.

Hens can produce eggs even without a rooster.

Feeding the Chickens

As soon as you have bought the chickens, it is time to find out what you can do best to nourish them. The best diet for these birds will always be what Mother Nature provides for them – worms and insects, growing grass, clover, weeds, buckwheat, and seeds. If you have a rooster, you might also catch it feeding some mice to the hens.

Take note that they also need to eat some grit or coarse dirt because this helps them grind their foraged foods.

Backyard chickens can eat your kitchen scraps except for raw potatoes, beans, garlic, anything citrus, and onions. These scraps can make eggs taste different. Be sure to check on the things that your chickens are not supposed to eat, such as the pine shavings that are supposed to act as the litter and Styrofoam.

Hens that are raised on pasture produce deep orange egg yolks and have bouncy, viscous egg whites.

If you cannot pasture the chickens, then at least have a run attached to their coop so that they will not be depressed.

There are also supplements to the main feed, such as oyster shells (provides calcium), cabbage (more of an anti-boredom food for them), and commercial feed.

Just in case you run out of feed for the chickens, no worries – you can hard boil and then chop some eggs and have them eat them. Chickens can even go for a day or two without feeding, but make sure that they have ample drinking water.

If you are feeling up for a challenge, you can also make feed mix out of the seeds and grains that you grow in your backyard.

The feed types also depend on the age of the chickens that you are feeding. Chick starter is fed to chicks during their first six weeks. This should be about 22-24 percent protein if you are raising meat birds (broiler starter).

There are medicated and un-medicated varieties of chicken starter.

The grower pullet is fed from seven weeks up until 14 weeks. This has to be rationed because the laying chickens need to be given a diet with lower protein while the meat producers need to be fed 18 percent protein.

After this, the pullet developer (finisher) is fed at 14 weeks till about 22 weeks. The feed stores often sell a grower-finisher that has just the right amount of protein.

Laying hens at 22 weeks will soon require 16-18 percent of protein plus some calcium and minerals. They need all these for stronger eggshells. Layer feed needs are strictly fed when the bird is 22 weeks old because younger birds get damaged kidneys.

Roosters have been observed to eat laying rations.

As for the broiler rations, these are high-protein feeds meant for meat birds. The Cornish X Rock grows quite fast at 18-20 percent protein in their diet. The protein content for the grower-finisher feed may be lowered if the broiler is reaching its butchering age (about 12 weeks).

Heritage birds (chickens that can retain their historic characteristics) may be kept on a higher protein diet until they are slaughtered.

The only way to decrease the amount of commercial feed that you give to your chickens is to allow them to pasture because, then, they will eat enough weeds, bugs, grasses, and seeds.

Building the Chicken Shelter

When the chicks have matured and are ready to move outdoors, you know that you already need a better shelter for them – but which one should you set up? Will it be a chicken coop, tractor, or henhouse? And here is another question that some homesteaders have already asked – can the doghouse or old shed be turned into a chicken coop?

The kind of coop that you build largely depends on how long the chickens are going to stay in that structure. Also, you need to plan whether you will construct a stationary and permanent structure or one that can be moved about.

Next, determine the amount of space that you are willing to set aside for your chickens. Never make the mistake of underestimating the measurements. You could begin small but might decide to expand the flock later.

Build a medium-sized coop as you test the waters.

If the backyard can provide ample foraging for your chickens, then you can construct two-three square feet of the coop for every bird. A bigger space is, of course, better because there will be more space for the chickens to roam around.

If you plan for the birds to be constantly cooped, or at least only during winter, then aim for about five to ten square feet for every chicken.

Chicken tractors can be moved around, and the right size for such is about five square feet for each chicken.

These are all general guidelines for backyard chickens. The bigger and more numerous the chickens are, the larger the space that you will need. Meat birds generally need bigger space than laying hens, just as full-grown pullets need a wider space than chicks.

Aggressiveness and pecking – should they ever become an issue – can be easily addressed by adding more space. Most of the time, these are just your chickens' way of saying that they are bored.

Chickens coops can be as simple as a wooden box with chicken wire. Save money by not going for the more elaborate designs. For urban and suburban folks, what you need is a secure coop that conforms to the building codes, homeowners association's requirements, zoning regulations, and other associations.

Laying hens typically need nest boxes. One box should be enough to accommodate four-five hens. Do not put in too many boxes because hens become broody (they constantly sit in the nest box) as they exert more effort to hatch the eggs.

Place the one-foot square boxes at two feet off the ground. Line them with shavings or straw, then attach them to the wall or place them on the shelves.

Laying chickens also need roosting space. The rule of thumb here is to provide six to ten inches of roosting space for every chicken. Again, the roosts must be two feet off the ground.

Check that the coop has ample shade, ventilation, and dust baths (dirt or dry soil). Place it on the eastern side of the homestead so that the chickens are warmed up during the morning but not burned in the afternoon.

A basic DIY portable chicken coop plan

Materials:

- Lumber
- Metal braces
- Screwdriver and screws

- Sandpaper
- Carpenter's square
- Power miter
- Table saw
- Metal hardware cloth

Method:

1. As soon as you have chosen the coop design, it is time to gather the 2'x 4' lumber for the bottom frame. Use metal braces for the corners and then screw the joints.

2. Make a perfect square with the bottom part of this frame. Use carpenter's square to check.

3. This movable coop will measure 9' x 8' or 72 square feet that is good enough for 24 chickens.

4. Rip the 2 x 4 lumber into thin pieces. These are going to be used for the diagonal rafters.

5. Using deck screws, assemble the pieces.

6. Traditionally, the space between the rafters is about 16", though 12"-14" can be good enough for smaller spaces.

7. Build the end frame using three vertical studs, forming a doorframe at both ends of the coop.

8. You can also attach two long stringers that will serve as the roost bars. Place a vertical bar to support the roost bars. Add these only when you are raising laying hens – no need for these when you are raising broilers.

9. You can also add nest boxes made of 2 x 2 boards secured with deck screws.

10. Place this box in the middle of the already constructed coop using deck screws.

11. Have a metal hardware cloth stapled at the bottom of the entire frame.

12. Add 4" to 6" of plywood and fasten with a deck screw.

13. Position the entire coop in your chosen location.

Caring for Chickens in Winter

Here is rule number one – chickens do not need their own heaters during winter. Now that is established, here are some tips that can help you raise chickens even in the cold of winter:

Deep Litter Method

This is a sustainable chicken manure management that many small homesteaders practice. It is a pile of compost (including your chickens' poop) in the middle of the coop flooring.

Just like any compost pile, start with a pine shavings layer, and then the chickens should take care of the poop. Add shavings, so the floor compost develops nicely. Don't worry about the aeration because the chickens will also take care of that for you.

The microbes in the litter should serve as probiotics (of sorts) for your hens.

Clean this annually or bi-annually, and you can then use the compost in your garden. For urban and suburban dwellers, you can simply clean the chicken litter weekly or monthly and then add it to your small compost bin.

Use Supplemental Light

Great layers such as the Buff Orpington will still lay eggs during winter, but, generally, supplemental light is needed. A 40-watt light bulb that hangs at 7' off the ground should provide the right light intensity for your chickens to remain healthy. This is enough for a small chicken coop, the size of 100 square feet. For bigger coops (200 square feet or more), buy a 60-watt light bulb.

If you are willing to forego the winter eggs, then you can just respect the chickens' natural cycle and wait for them to be active again. Hens, after all, naturally rest during fall as they experience molting (losing their feathers), and their egg production decreases.

Feed Them Corn

Feed the chickens some cracked corn in the evening so that they will stay warm. Corn is one of their favorite foods, so they will be happier as they rest with full tummies.

Hang a Cabbage Head

You wouldn't want your chickens to be bored and depressed during winter, so you need to give them something to keep them busy. A simple cabbage head hanging on a string could be their toy. All the pecking should keep them busy all winter.

Use Petroleum Jelly as Anti-Frostbite

During the coldest winters, you can keep the bigger breeds with huge combs and wattles safe from frostbite by applying petroleum jelly on those parts.

All in all, just let the chickens do what they want during winter. If they want to stay longer in the coop, then allow them. And they are a hardy bunch, so there is no need to shoo them indoors when it rains or snows.

Predator Protection

Potential chicken predators are:

- Dogs and cats (even your own pets)
- Hawks
- Foxes
- Weasels, minks, ermine
- Coyotes
- Raccoons
- Bobcats
- Snakes (they love chicks)
- Owls
- Fisher cats

- Rats

To keep the coop safe, you can dig a trench all around it (about 12 inches deep). Bury hardware cloth so that the digging predators cannot go in.

You can also elevate the coop so that rats, mice, and weasels will not get into it. Check the bottom of the chicken coop and be sure to patch holes. Turn on some form of lighting at night, or you can have motion sensor lights installed.

The compost pile must be kept at a good distance from the coop and make sure that feed leftovers are cleaned by nightfall. Use hawk netting for owl and hawk problems.

The last layer of protection for your livestock is a gun. Firing in the direction of the predator should be enough to deter its entry. Traps can also offend most predators, but be sure only to use these last two defenses if you have already tried the simpler ones.

Your Homestead Pantry

So you have learned to grow and harvest your food. And now the fruits and vegetables – all colorful and beautiful – sit in baskets waiting for what needs to be done. Homesteading has, indeed, giving you fresh provisions that could potentially sustain your family.

Having year-round food security is the next step for your homesteading life, and you will need a homestead pantry to do that. Many people use pantries for smaller amounts of food, but as a homesteader, you could stock it up on months' worth of food (your goal is to build an annual stock for you and your family).

Do not force yourself to fill up huge shelves and freezers straight away – it takes time to transition from your current practice of grab-then-go to a life of stocking pantries.

To make things easy, you can divide pantry essentials into two groups: foods with a longer shelf life (e.g., rice, flours, and olive oil), and the perishables (e.g., meats, dairy, and fresh produce).

The Essentials

Flours

You can keep the all-purpose staple flour handy, but with the unhealthy gluten that you could get from it, perhaps, it is time to consider healthier options, such as whole grains, sourdough, and einkorn.

You only need to replenish flour every few months. Make sure also that they are stored in bins that are not directly exposed to sunlight. Check the flours every few weeks to make sure that they still smell fresh with zero traces of rancidity.

You can keep stocks of tapioca flours and whole-wheat flour, too.

Fats and Oils

Exclusive of butter, you should store the essentials:

- Lard
- Avocado oil
- Olive oil
- Coconut oil

You can also include a bottle or two of extra virgin olive oil for those delicious salads and sauces that you'll soon make. This will also be a mouthwatering dip when paired with basil (pesto dressing).

Sweeteners

It is time to cut down on your refined sugar consumption. Use natural sweeteners that will not give you the same sweet cravings as sugar does. There's maple syrup or honey that can be healthier alternatives. Organic cane sugar is also a healthier choice.

Additional Items

- Homemade brown sugar
- Turbinado (raw) sugar

- Molasses

Legumes, Dried Beans, Nuts, and Seeds

Every homestead pantry is not without dried beans. These can provide vitamin B, iron, minerals, selenium, potassium, calcium, and magnesium. They are also perfect ingredients for winter soups.

Keep stocks of seeds like split pea, lentils, and barley since these are also great for soup recipes. Other dried beans that you can harvest from your garden and include in your pantry are:

- Black beans
- Kidney beans
- Cranberry beans
- Pinto beans
- Popcorn
- Cannellini beans

Nuts can also be stored for as long as six months. So, go ahead and keep containers of salted almonds, walnuts, pecans, peanuts, and sunflower seeds.

Canned Fruits and Vegetables

Your list of canned goods could change as the season changes. This depends largely on what crops you have harvested during the season. Canning, pickling, and other methods can preserve in-season vegetables, but be sure to do the process while the vegetables and fruits are still fresh and not when they are about to go bad.

As soon as you harvest these staples, make a habit of preserving them so you can ration your fruits and vegetables throughout the year:

- Tomatoes (this staple food preserves)
- Sauerkraut
- Green beans
- Beets

- Applesauce
- Peaches
- Cucumber
- Olives
- Corn

Spices and herbs can also be preserved when you add olive oil and then pour the mixture in ice trays.

Dried Herbs

You can enjoy fresh herbs from your garden, but it would also be wise to store some by making dried herbs.

Keep these herbs on hand:

- Thyme
- Bay leaves
- Oregano
- Tarragon
- Rosemary
- Herbs de Provence

Vinegar

Just like olive oil, vinegar can help change the flavor of any dish. Vinegar is pretty easy to make. You can use leftover fruit juice, wine, or any alcoholic beverage. Here are a few recipes that you can use to make your own vinegar.

For white vinegar –

Ingredients:

Half a gallon of water

1-¾ cups of sugar

A packet of wine yeast or baking yeast

Balloon

Two cups of unfiltered vinegar

Method:

1. Begin making the mother of vinegar by combining water and sugar. Heat them in a large pot, then keep stirring until the sugar has dissolved. Allow the sugar water to reach 110°F.

2. Pour the mixture into a glass jug then stir in the wine yeast.

3. Form an airlock by placing the balloon on the mouth of the jug. If the balloon becomes too big, make sure that you lift the neck to release some air.

4. As soon as the bubbles are gone (about two weeks), you can already pour the liquid into a stainless steel or huge glass container.

5. Put the unfiltered vinegar into the container and then cover it with layers of cheesecloth. Seal the neck using a rubber band.

6. Put the jug in a dark room at room temperature.

7. The mother of vinegar will form after a few days. This is that thin film over the vinegar surface.

To make other vinegar recipes, you can just replace the sugar water mixture with hard cider, leftover beers, white wine, red wine, or just about any alcoholic beverage.

Salts

You could switch to sea salts for your organic recipes or use them to preserve your food harvests.

Keep these three salt types on hand:

• Himalayan sea salt

• Celtic sea salt

• Fleur de sel

The pink Himalayan sea salt is great for everyday seasoning, the Celtic sea salt has high mineral content, and the fleur de sel flakes are great toppings for bread.

Declutter, then Organize

The rule here is simple – toss the things that you have not been using for more than six months (except your winter clothing and accessories) and then keep the ones that you use daily or weekly.

Next, put the dry goods into clear containers. You want to see where the sugar or flour is with one glance. Group the grains, beans, flours, sugars, and other pantry essentials. The bulk food items should be placed in the pantry, while smaller jars can be placed in the kitchen where they are readily accessible.

This basic homestead pantry can be expanded as you grow your homestead. Just take your time, plan, and soon you will save a lot of money as you reduce your grocery shopping trips. If you religiously preserve your harvests, then, eventually, you might even be able to feed off the land one hundred percent.

PART THREE:
HOMESTEADING SURVIVAL

Extreme Weather Conditions on the Homestead

Do you know how some homesteaders are also doomsday preppers? They can't be blamed because the conditions on their homestead – the extreme weather conditions that can change in a snap of a finger – could easily make anyone feel fearful. But this need not be the scenario for you, too.

Earth's weather does bless (and sometimes curse) farms and many homesteads throughout the globe. Most of these conditions are manageable, but you can prepare for adverse weather.

Drought

A drought is an event where there is a prolonged water shortage. With a family that would want to depend on homesteading, this is, indeed, bad news. There are many ways that you can prepare for drought, and these are broken down here according to your home, the garden, and your livestock.

Your Home

You can already use the rainwater barrel systems that you previously set up. Assuming that no laws are prohibiting you from setting up the rainwater collectors, then you can fill them up and store water.

Graywater (used, soapy water) can also be stored for toilet cleaning. Just make sure that you do not use harsh detergent.

Your Garden

You can also use the stored rainwater to water your plants. Mulching the garden will also benefit the plants, but stop fertilizing while the weather is not improving.

Pull the weeds so that the plants will not have nutrient competitors.

The Livestock

Make sure that all of the animals have a steady source of drinking water. Monitor them daily so you can take note of any that have grown weak. If there is a lactating cow or goat that is not currently able to provide milk to its calf or kid, then the offspring must be given supplements and feed.

Care must also be taken to keep animals from being trapped in dried up dams.

Life-Threatening Heat

There is a phenomenon called the El Niño, which is periodic, large-scale warming that happens in certain areas of the planet. It can get extremely hot, so the effects of this event could affect you, your livestock, and the plants. Or there are moments when the summer season is just tremendously hot.

No one is immune from its effects, so it is best to protect yourself and every living creature on your homestead.

Your Home

Check your homestead pantry and see if you have ample stocks for your family's consumption. Make sure also that you have ample drinking water for everyone.

Dry, hot seasons aren't all bad, though. You can use the weather condition to dry fruits and vegetables, and even some herbs that you will store in your pantry.

Your Garden

Install a shade cloth over the garden beds where heat-sensitive crops (e.g., eggplants, tomatoes, peppers, and cucumbers) are planted. You can also plant these crops during the cool months so that you don't have to worry about losing them during the hot months.

See the drought tips on how to keep the soil moist.

Early on in your homesteading years, you can also make sure that you plant deciduous trees that can eventually cast a shade over your garden.

The Livestock

See the Drought section for livestock care tips.

Snow

If there is too much heat, then your homestead could also potentially suffer from extreme cold. There are ways to protect your home and your garden during winter.

Your Home

Install insulators in your home so that you, your family, and pets can all snuggle safely indoors.

Your Garden

You can also install insulators in the greenhouse or heat it if the coldness goes way below the freezing point. Be sure to remove heavy snow from the hoop houses, greenhouse, and row covers.

Use row covers for the garden beds, and while the sun isn't showing, you can plant kale, leek, collard, cabbage, rutabaga, Brussels sprouts, turnips, parsnips, beets, and carrots.

Yet again, do heavy mulching so that the root crops will not freeze.

Winter is also the best time to organize the seeds and plan for poultry additions. You could even take this time to explore new plants and herbs that you could eventually plant.

The Livestock

You already learned how to take care of chickens during winter. As for the other animals, your goal while heavy snow is out there should be to keep them busy and happy.

Keep all animals well-fed, comfortable, and be sure to assist any pregnant animal. If you pasture most of your livestock, then they can be taken to the closed barn during winter. Just make sure that there is ample ventilation, food, and water.

The living areas also need to be cleaned regularly with the manure taken out. There should be enough space for the animals to walk around so that they can still exercise.

Meet every animal's nutritional requirements. Refer to the previous chapters for their feed and supplement needs.

Wind

Your Home

Winds can come in different forms – like a hurricane, windstorm, tornadic winds, even gusts. To care for all your loved ones and even the structures on your homestead during windy seasons, be sure to stay indoors and go to the sturdiest shelter. If emergency managers are already telling everyone to evacuate, then leave – immediately.

Your Garden

The strongest gusty winds stress the plants during storm seasons.

Be wary of breaking branches and stems when the storms wreak havoc. Two combinations of weather could occur.

First, it can be extremely hot and windy. Even the slightest breezes can affect seedlings, so you need to protect your garden when these winds occur.

Second, it could get chilly and windy. The wind could heighten the effects of cold weather. Use a burlap windbreak during these wintry windy days. Support this setup with stakes. You can also create a wire mesh or ring for each plant and then fill it up with oak leaves or straw.

Don't forget to mulch so that the heat or cold effects are decreased.

The Livestock

If they cannot be evacuated, then the animals should be confined to the sturdiest shelter on your homestead. Go back to them the moment the situation clears.

Lightning Storms

Your Home

Do keep in mind that it is never a good idea to go to the garden when a lightning storm is happening. Stay indoors at all times, and wait for the storm to stop. As soon as it is over, that's the only time that you should check on your garden, the chicken coops, and barn.

Your Garden

Keep all sensitive electronic gadgets away from the garden when there is a lightning storm forecast. The greenhouse controllers and irrigation timers should also be brought indoors.

The trees must also be trimmed so that they won't be damaged.

The Livestock

Install lightning protection systems at animal enclosures. Next, be sure to put fences all around the solitary, tall trees. You also need to lead all the animals away from water sources.

All animal shelters must also be built away from powerlines and trees.

Living with Homestead Pets

Deciding to move to a homestead and alter your lifestyle is a huge step to make. By now, you know that it has benefitted your family in terms of health and perspectives; also, the new adversities that you experienced might even have strengthened you.

But have you ever wondered how your pets have also adjusted to this decision? If you already have dogs and cats before homesteading, then you need to get them to adjust to your new lifestyle. No more of those lazing-around-on-the-couch back in the city – it is time to welcome everyone to rural living.

It is all about ensuring that your pets still get the best environment even when they are already rural animals. Just as there are no pizza deliveries for your family, you will also have limited access to dog food, cat food, cat litter, and other furry pet needs, so adjustments have to be made.

You can choose to keep the pets indoors but think of all the advantages of training your dog to help you with your homesteading chores. Unsurprisingly, training your current dog to become a farm pet is just like getting a new farm dog and introducing it to farm life. Cats, though, are a different story.

The Farm-Friendly Dog

Every homestead is unique, but there are common aspects, such as open space, gardening, chickens, other small animals, and predators.

Choose the dog that you will have on your homestead. Livestock guard dog breeds are:

- Central Asia Shepherd
- Akbash
- Maremma Sheepdog
- Anatolian Shepherd
- Armenian Gampr
- Polish Tatra
- The Great Pyrenees
- Kangal
- Tibetan Mastiff
- Russian Ovcharka, and
- Sarplaninac.

Have about eight well-bonded dogs form a pack. Their age could range from one and a half years to 13 and a half years. These dogs can help keep predators away from the homestead so you won't have to worry too much about foxes, stray dogs, eagles, hawks, and wolves. Certain farms even have mountain lions as predators, so be sure to have a pack of dogs ready to defend you.

Powerful, large dogs can also help pull the sled when you gather firewood and haul barrels of drinking water. They can ease your workload while providing personal protection.

Akbash and Great Pyrenees breeds are also great sources of fur. Their fur can be spun into yarn because they are as soft as baby alpaca fur.

Since these farm dogs are giving back the kindness that you are showering them, you should also make sure that they remain healthy at all times. Get in touch with a veterinarian for your dog's vaccinations. The five-way vaccine that counters kennel cough, hepatitis, parainfluenza, canine distemper, and parvovirus should be administered. You can include the anti-corona virus and leptospirosis if you opt for the seven-way vaccines.

Six-month-old puppies should already have their rabies vaccination.

As for their diet, you need to check the fat and protein content of their food. Growing pups need 22-25 percent protein, while adult dogs need only ten to 14 percent. Pregnant and nursing dogs need more protein (20-30 percent).

As for their fat needs, growing pups require just eight to 20 percent depending on their body mass. Adult dogs need five to 15 percent while pregnant, and nursing dogs require ten to 25 percent.

Carbohydrates are a great energy source for dogs, so you can include rice, oats, barley, and other grains in their diet. Do not feed them with human foods because this could lead to health issues and even early death.

If your farm animals can survive with kitchen scraps, well, your dogs must not. Add vitamin and mineral supplements, too.

The Barn Cat

Again, you need to decide if you are going to keep a house cat or a barn cat. Here are some words of caution when it comes to cats in rural settings:

Never place city cats in a barn.

Never put kittens in a barn.

Kittens do not have the survival skills needed to evade horse hooves and such. Feral kittens, though, could survive.

Remember also that barn cats are more susceptible to parasites, rabies, feline leukemia, predators, and being hit by trucks. They can, however, help in controlling the rodent population, so there would be decreased grain losses and farm equipment damage.

An ideal barn cat is one that came with the farm, meaning it is a stray kitten that adapted to your family as you moved in. Just about any breed of a cat could live on a farm, but certain breeds could have advantages.

Shorthaired breeds will not let you worry about matted hair, hairballs, etc. Female cats are naturally better hunters compared to males, and the orange male variety is known to be the sweetest.

You need to confine a cat first if you want it to stay in your barn. Cage an old cat for about two-four weeks while new cats can be freed after a month. Include a litter box in the cage and feed it frequently so it will associate the cage with a safe zone.

Barn cats should still be fed even when they catch mice on your homestead. Feed them twice daily and provide ample fresh water. During winter, make sure that their water bowls are also heated. Schedule the second feeding at night and make sure that it is canned food so that the cats will stay indoors. As you do so, you keep them safe from predators like owls, coyotes, and raccoons.

Cats also need a lot of protein, so feed them with fresh fish and cooked meat.

Deworm the cats twice each year and make sure that they have a lofty station to rest on. You can also build their loft inside a sturdy shelter so that they have somewhere to go to when the weather isn't too friendly.

Just like dogs, barn cats also should be administered vaccinations. Combination vaccines for adult cats are the FVRCP that can counter feline distemper, feline leukemia, FeLV, calicivirus, and rhinotracheitis.

Treat your barn cats and farm dogs as members of your family by providing for their needs and loving them. They will even help you with some homesteading chores, so you should appreciate them more.

Homesteading Alone

Most homesteaders take on the homesteading journey with a companion or their whole family. If you begin this journey alone, then you have more things to consider. It is not easy to homestead even when you are a pack, but taking on this lifestyle alone, equates to a lot of preparation. Failure to prepare could lead to sure disappointment.

The upside to homesteading alone includes having a sense of accomplishment. Imagine being able to grow and harvest everything that you need (on your own), protecting yourself, your plants, and livestock, and you are your own boss.

You will learn to be crafty when you are homesteading solo. You simply have to find ways to solve issues, and you can make things better without getting anyone's aid.

Sometimes, the scenario is that you still live with your family, but you run the homestead alone. This is okay, too.

Here are some basic projects that you can accomplish on your own:

- Build chicken coop
- Build other animal pens
- Make raised gardens

- Plant vegetables and fruits

- Build a fence all around the homestead

- Design and implement a compost pile system

- House maintenance

It is not impossible to do these even when you are alone, but you need to learn strategies that can help decrease the workload, avoid injuries, and take care of just the right number of animals.

Game-Changing Decisions

Reconsider the Number of Livestock

Homesteaders who love to take care of chickens happen to commit the same mistake – buy more of them. And before long, they see their farm teeming with chicks that are more than they can handle.

Or you might have taken up the egg production business, where you sell organic eggs, and before you know it, your success is eating up most of your time. Delivering 60 dozen eggs each month is a huge task for a lone homesteader.

Also, you need to feed the chickens, so the more that you keep, the more time it requires to feed them properly.

Chickens can forage for insects, nuts, and berries, but feeding them during the winter season can also be a challenge.

You will come to a point when you need to decide whether you will just harvest fruits, vegetables, and meats for your family's consumption. If you are literally alone, then see if you can just harvest for your needs.

You have to take care of yourself, too, so learn to cut down the chores and your livestock duties. Downsizing could give you that much-needed sanity, so take the time to consider it.

Purchase the Right Equipment

It is also important to find, buy, and maintain the right equipment. You do not have to buy more because, for solo homesteaders, one set of equipment could address all needs.

Don't let equipment care eat your time. Buy quality equipment the first time so that you won't have to keep fixing it. Invest in a wheelbarrow, pull wagon, and basic gardening tools, like a hand trowel, garden scissors, weeder, pruning shears, curved blade shovel for digging, and flat digging shovel.

Reassess Your Goals

Zero in on the goals that matter to you this year. List the things that bring you enjoyment, and be sure to include those as your goals. If keeping goats and fowls while tending to a garden already stresses you, then why do it?

Decrease the pressure on your workload, and you will surely lessen the stress on your body.

Know the Downsides

Homesteading has many upsides, but there are also some downsides to it. To begin with, there are limitations to being a solo homesteader. Heavy lifting is almost impossible to do on your own. There are just a lot of homesteading duties that need more people.

Homesteading alone also brings just your skillsets. If your homestead suffers from an electrical problem because of a recent lightning storm and you have zero knowledge about electricity, then you will pay a professional to do it. This is also the case with repairs on machinery and equipment.

While the joys of homesteading include being proud of your solo accomplishments, the downside to this is that you also have no one to share your problems with. And homesteading can be overwhelming so it could get lonely.

Sustaining injuries or getting ill could also be glaring disadvantages to homesteading on your own. While you can call 911 and wait for a response, what about for situations that require, say, CPR?

Battling Loneliness

Face it – homesteading alone can get flat-out sad. The isolation will not make quick trips possible, so you are alone in the middle of nowhere.

Homesteading solo will result in more alone time. You might feel then that the walls are caving in and that your decision to homestead might be a mistake.

Many homesteaders have admitted their struggles with rural melancholy, but some fight it with chores on the farm. As you previously learned, you could be an introvert or extrovert, so, at times, it is all a matter of personality.

Introverts might survive longer in isolation, but extroverts need to divert their thoughts to other things. If you feel that socialization should be a part of your day-to-day existence, then do not sign up immediately to a solo homesteading life.

Think twice before changing your lifestyle.

Now, if you have already made the decision and you need to cope with slower days, then here are some simple tips:

Start a Blog

One bored person + Lots of Time + the Internet = Probable Passion

You can begin an online business to keep busy or write a blog. Dig deep within you and find out what really excites you. If you want to blog about your decision to homestead solo, then go ahead and write about that. With homesteading becoming more and more popular each year, all it takes is consistency in writing, and you could become an Internet sensation.

Think That It's a Transition Period

The first part of homesteading is always the hardest – every homesteader can attest to this. If you look at the first few months as an adjustment phase, then you can prepare and plan for the upcoming months.

There is so much to do on your homestead, so do those things. As you get used to doing them, then you will soon have a renewed passion for homesteading.

Socialize

Use the Internet to seek out your old high school friends on Facebook. Email those that do not have Instagram or Facebook. You don't need physically to be in front of a person to socialize.

Nowadays, a single click of a button means you can get in touch with your relatives or friends.

Homesteading alone need not be lonely. It is up to you to make it more exciting and fruitful (literally and figuratively!).

Money-Saving Crafts and DIY Projects

You have just learned about how homesteading can get quite lonely, at first, as well as how to counter melancholy on your homestead. Another way to counter rural loneliness – and an effective one at that – is to make some DIY crafts.

Instead of spending a hefty amount on furniture and appliances, how about you craft your own? If this is too big a project for now, then try these simple DIY crafts:

DIY Aqua Dam

If you set up your homestead in a flood-prone area, then you need to make sure that your family will be protected from the onslaught of rains. Building an aqua dam should solve this problem for you.

An aqua dam is a barrier filled with water that can help divert waters. All you need are some inner tubes, fill them with water, and then create a temporary barrier.

Organic Weed Killers

Support and protect the garden from invasive species. The easiest solution is to use vinegar at its full strength. Pour apple cider vinegar into a sprayer and spray directly on the affected areas. A word of caution – it could also kill good plants, so spray only when you are sure that it is the weed that you're killing.

This is a pet and livestock safe weeder, so spray away. If you are still worried, then you may reduce its acidity by five percent. Just add a little water.

Other Options

Vinegar and dishwashing soap mixture

Use an ounce of dishwashing soap with every gallon of pure vinegar. This can also be used as a garden insecticide. If you do not want to harm the plants, then use organic dishwashing soap as well.

Vinegar + soap + salt

Use a gallon of vinegar to every cup of salt and one tablespoon of dishwashing liquid. Mix the ingredients then apply to the affected areas.

Vinegar + lemon juice

Many homesteaders can attest that mixing lemon juice with vinegar can heighten its weed-killing properties. Add a cup of lemon to a gallon of vinegar.

Vinegar + essential oil

Mix pure vinegar with a tablespoon of orange or clove essential oil. The oil is believed to stick to the garden plants, thus, increasing the mixture's effectiveness.

DIY Backyard Incinerator

Also known as a burn barrel, this is a 55-gallon metal drum with an open head. You can modify this to burn household trash cleanly and safely.

Materials:

- Steel drum
- Drill or handgun to make the air holes
- Two concrete blocks
- Heavy fencing or metal grate (to cover the mouth of the barrel)
- Sheet metal (as barrel cover)

Method:

1. First, pick a location for your burn barrel. This should be away from trees and not where it could encounter prevailing winds. Situate it away from any combustible material.

2. Drill air holes into the metal drum; about 20 on different sides and at different heights.

3. Drill four more holes, measuring about half an inch each. These should be drilled at the bottom of the metal drum so that rainwater can drain.

4. Use a fire cover (grill top) and bend it so it can loosely fit.

5. The steel siding can serve as the rain cover for the burn barrel.

6. As soon as the holes are drilled, set up the barrel on two concrete blocks. The edges of your drum barrel should be placed on the blocks, but air space must also be left. This allows air to flow freely and helps with the drainage.

7. Cover the barrel when it is not in use so that the contents will not get wet.

DIY Citrus All-Purpose Cleaner

Many people admit that they are tired of paying big bucks for chemical cleaners that require a gas mask to use safely. The thing is, homemade versions are just as good – if not better.

If you have been using a 50-50 mixture of vinegar and water for the longest time and you want to stir things up a bit, then you can try this citrus cleaner for a change.

This citrus cleaner is still an organic mixture that you can use without worry. It is even a wonderful way to use the citrus peels that the goats and chickens don't pay attention to.

Ingredients:

- One to two quart-sized mason jars
- Spray bottle
- Strainer
- Lime, orange, lemon, or grapefruit peels (two or more of these would also work)
- Three-four drops of grapefruit, lemon, or orange essential oils (optional)
- Water

Method:

1. Fill the jars up halfway with citrus peels. Pack the jars if you want to. You can also combine grapefruit, lemon, and orange peels.

2. Fill the rest of the jar with white vinegar. Cover then give it a good shake.

3. Steep the vinegar and peel mixture for two-three weeks. The longer that you leave it, the more potent it will become.

4. After the required steeping period, take out the peels and strain using a fine strainer. This should remove the citrus bits that are left floating in the mixture.

5. Dilute with water (one part mixture with one part water) then pour it into a spray bottle.

6. Add three-four drops of citrus essential oils to boost the scent further.

You can use this natural cleaner on the floors, bathtub, kitchen sink, countertops, and toilets. Have fun cleaning because you don't have to suffer from the chemical odors any longer.

DIY Paper

Making basic paper is easy and fast, but it can take several days before it dries completely.

Method:

1. Use a food processor or blender to shred the ingredients. You can use newspaper (this is the most economical) to start. Add just enough water so that the paper becomes damp but not wet. Add a glob of white glue as well.

2. Spread the blob on the fine screen. Flatten and smoothen the paper.

3. Put it in a dry area, then wait for the paper to dry. Never place in direct sunlight, though. If mold starts to grow, then you need to throw it out and restart.

Homemade paper can be used as gift tags, Origami material, and journal sheets.

DIY Candles

Do you like the sweet scent of beeswax candles? Homemade candles are much safer for you because they do not release toxic matter into the air. There are three ways that you can make homemade candles:

Candle Dipping

Hold the hot wax in one container and cold water in a second container. Take turns in dipping into the hot wax, then into the cold water, and into the hot wax again.

You can make two candles at a time when you use both ends of a wicked string. Hang the candles to dry for a couple of days.

Candle Molding

You can also get creative and craft the shapes that you want. You could start with the basic empty milk carton. Place ice cubes around wicks then pour in the hot wax.

Allow the mold to cool before removing the carton. You will see that the candle will have beautiful, decorative holes.

Rolled Candle

Another fun way to make candles is to roll a wax sheet around the center wick. Make beeswax or buy it to create a honeycomb mold.

Lay the wick along the edge then roll it up. Warm the beeswax sheet first in your hands so that it won't crack.

Here is a little safety advice – never leave burning candles unattended. Be sure to create and discuss a fire safety plan in your home. Install fire extinguishers, smoke detectors, and other fire deterrents.

DIY Detergent

Homemade soaps and laundry detergents are catching a lot of attention lately because of their potential money-saving aspects. This solution that you are going to learn now can clean even the toughest stains (e.g., stained cloth diapers). Compared to commercial detergents, it is even better because it does not contain phosphates and sodium lauryl sulfate.

Ingredients/Tools:

- One bar of soap (about $3.25)
- Half a box of borax ($3.97 per 72-oz. box)
- One cup of washing soda – not baking soda! ($8.49 per 55-oz. box)
- Several cups of water
- Essential oils (different price for every scent)
- Five-gallon bucket
- Water
- Cheese grater
- Medium saucepan

Method:

1. Using a cheese grater, coarsely grate the soap.

2. Put the shreds into a medium saucepan filled with a few cups of water.

3. Heat then stir until all the soap pieces dissolve.

4. Add borax and the washing soda. Stir and dissolve the recently added ingredients.

5. Pour the mixture into the five-gallon bucket. Fill it three-quarters of the way with hot water. Stir thoroughly, then leave it to sit, undisturbed, throughout the night.

The homemade detergent will look like a gel chunk, and its yield will be about three-four gallons depending on the amount of water that you added. You can leave the block of detergent in the bucket or use smaller jugs or buckets to store the smaller portions.

You can dispense using a drink dispenser, or you can just scoop it out of the bucket when you finally need it.

DIY Curtains

Learning how to make your own curtains can give you the freedom to make shapes and designs that you want. Isn't it just frustrating if you find the right color, patterns, and texture on a curtain only to be told that it does not come in your window size?

Homemade curtains are also better for people with allergies because now you can choose the materials to use. Find curtain materials that are easily washable every couple of months. This is crucial for people who suffer from dust mite allergy.

You will need basic sewing skills to do this project. If you are good with those, then read on:

Method:

1. Measure the windows that you want to cover. Remember to measure beyond the window frames.

2. Create dimensions for every layer. Add seam allowances of about 5/8 (standard).

3. Decide how each of the curtains will be attached to the rod. The hanging distance will shorten with the depth that will wrap around the rod, so be sure to include that in your calculations.

4. Measure the amount of fabric that you need for this project.

5. Draw the patterns on a big piece of tissue paper. As long as your design is symmetrical, you can fold the fabric as you cut along its edge.

6. Continue with the sewing pattern. Pin the paper, cut, and then sew.

DIY Baskets

Baskets have many uses on a homestead. Imagine carrying the basket around and filling it with freshly harvested fruits and vegetables. There are various basket-weaving techniques that you can employ.

You can make wisteria baskets or be more creative with the bark basket. Poplar trees undergo a growth phase from April until June. In this phase, the bark can be easily cut and peeled off.

You can use these peeled barks as the sides of your basket. Don't expect the basket to form a flat bottom because it won't, but you can create a strap for it.

Using a crisscross pattern, use hickory soaked in water as the lace. Punch holes on the sides of the bark, then lace them up.

The size of the basket will depend on the size of the bark that you cut.

DIY Paint

The egg tempera paint has been around for many centuries. Medieval paintings used a lot of this type of paint. Oil paint may also be formulated using linseed oil and milk paint with casein (dried milk protein).

Pigment (use natural materials like herbs, rust, and mud)

Binder (holds the pigment onto the painted surface)

Solvent (used to thin out paint)

Method:

If you want to make egg tempera paint, you can use watercolor tube paints, egg yolk as a binder, some water, and solvent (optional).

For the casein paint, use skim milk made sour with lemon juice. Leave overnight, then strain the curds and add the pigments. Don't worry about the odor because it will quickly fade as the paint dries.

Painting does not have to leave toxic fumes, and now you can keep busy when you do small painting projects.

Now, you can say bye-bye to boredom!

14 Homesteading Income Ideas

The DIY crafts that you just learned are potential cash cows. So, yes, you can sell the detergent, all-purpose citrus cleaner, curtains, and even the candles that you crafted. There are many online sellers of these crafts and organic cleaners, so you might want to follow their footsteps by creating a website for your unique products. However, since you said sayonara to a life of convenience to live a healthier (and potentially happier) life, you can look deeper into your goals.

Being a homesteader means you want a life outside of the easy life that you were/are currently living. So get back to that target and decide what you can get out of homesteading.

Sell Homegrown Products

Before you tackle this head-on, check the food safety laws in your area first. Once you have those covered, especially when you are going to sell milk or meat, then you can go ahead and sell your organic produce.

1. If you don't want to deal with the red tape, then the workaround is to sell the animals instead of their food products. This means you can sell the live chicken and not just its meat.

2. You can also sell chicken eggs or eggs from geese or ducks. One dozen chicken eggs could sell for $4 or more, so imagine if you can

sell 70 dozen each month (and this is still a small-scale egg business) – you get $280.

Everybody seems to want organic eggs nowadays. People prefer it for its omega-3 fatty acids that are good for your eyes, heart, and skin.

3. You can also sell gallons of fresh milk. The extra milk that your family can no longer consume may be sold in farmers' markets. You can also begin a goat-share or cow-share program if you don't want to worry about those laws.

A cow-share or herd-share agreement is much like owning stocks in a company. A consumer will pay a farmer for caring for his or her goat or cow and eventually for milking the animal. The owner of the herd-share will obtain the milk from his or her own animal.

4. Selling homemade dairy products, such as cheese and butter, can also be a source of income for you and your family.

5. You can also sell broiler meats and other fowl meats. For bigger sales, you can sell pork and beef.

6. How about selling heritage ducks, geese, or turkeys? Livestock conservation has also been causing quite a stir lately. Rare breeds of poultry and livestock are now being conserved and sold as heritage animals.

Consider the long-term plan for conservation breeding. What species would you raise – are they going to be large or small animals? Will you deal with fur or feathers?

Just as you would care for regular livestock, you need to learn about animal husbandry, feeding, and other animal needs. More importantly, you can add learning about the characteristics of the species that you will sell as a heritage animal.

7. How about fertile eggs? If you've got a rooster in the flock, then you can sell fertilized eggs online. See if you can have the fertile eggs shipped, but before you do, check the fertilization rates. Provide a rough estimate for the hatchlings because even the healthiest clutches seldom hatch out.

8. Have you heard of stud services? If you have ample space to keep the male animals, then you can rent out their service. This could make you some good money, especially when you have rare animal breeds on your homestead.

Keep tabs on the fertility of the roosters, the buck, and the male cattle. Even your farm dogs can be rented out for their stud services. If you get more clients from this business, then you might like to set up a stud farm.

You need to design a program that mitigates diseases and pathogens. Take time to set up a meeting with stud managers. You are the boss' boss, so you get to decide what is best for them; however, be sure to quarantine the ladies at some point and separate the pens of every male animal.

Register the sires, and you will bring in even more money. You might also consider semen collection done by certified or registered semen collectors.

9. Fiber animals can also provide income opportunities if extra activities such as shearing fleece are okay for you. The fleece that is harvested needs to be skirted and cleaned before you can market it. There is also a market for unwashed fleece.

Clean fibers are sold by the pound, and these can also be made into roving. If you have spinning skills, then you can spin a yarn of varying thicknesses and weights.

10. Selling honey can also provide a sizeable input into your homestead income. It's time to put your honey harvesting skills to good use. Harvest the raw honey, filter it, and then bottle it. The bees can take care of the honey-making process for you while you search for people who have a constant requirement for your product.

Probable honey clients are bakers, chefs, and those who make organic soaps, shampoos, and body washes.

11. Apart from the livestock sales, you can also sell plant products such as heirloom seeds (it has the same concept as the heirloom livestock) and transplants.

Transplants are extra plants that can be sold for ornamental or conservation purposes. Selling edibles such as berries, mushrooms, and fruits is also a great way to make extra money from your homestead.

12. Another profit booster is growing trees. Some homesteaders do not want to go through buying the seeds, planting them, and then growing the process of the tree.

As a tree grower, you would need fruit seeds to jump-start your business. The most profitable fruit trees that homesteaders search for are:

- Avocado – sells for $10/7" plant; sells at $30 for every 8" plant
- Mandarin – $5 per 7" plant; can be sold for $25 for the 8" plant
- Lemon – $5; $15
- Persimmons – $10; $35
- Orange – $5; $25
- Pomegranate – sells for $10 for every 7" plant; $35 for each 8" plant

Here's an easy tree-planting schedule

- One month – plant the fruit seeds inside pop cans.
- Two months – replant the one-two cm seedlings into plastic cups.
- Three months – replant the small plants into individual two-liter plastic bottles.

Since you grow the fruits yourself, you can go ahead and label the seeds as organic. If you can wait until the plant reaches two feet, then you can earn twice as much. Graft a branch from the same kind of tree. This should cause the tree to produce fruits.

You can grow seeds from any windowsill in your home. It is better to grow them inside because they can be exposed to torrential rains and snow. They could even be blown over by gusty winds, so have them grow in the safety of your home for now.

13. Make and Sell Furniture

One of the businesses that you can link with your homesteading life is crafting handmade furniture and selling your wooden arts. If you are truly serious about adding this to your load, here are the things that you need to do:

1) Find a niche that will suit your carpentry skills. There are home furnishings, office furnishings, custom-built pieces, and built-in cabinetry. You can sell your furniture online or have a small brick-and-mortar store within your homestead. You can even target a smaller niche, such as installing upholstery.

2) Location is everything, so your furniture-buying customers will surely be amazed to find your shop on a rural farm.

3) Obtain the relevant permits and licenses.

4) Write your business plan.

5) Don't forget to find inspiration for your furniture designs.

6) Lastly, create a handmade furniture website.

You can make furniture from different materials. You could learn woodworking skills or use metals or glass to create home accents.

14. Consulting Business

Once you become great at your homesteading life, you can even begin a consulting business. As a consultant, you can advise on how to start homesteading, what livestock would go best in what size and type of land, and what fruits and vegetables to grow. People can also ask for your expertise when they want to sell products from their homestead and when they want to expand.

Today, you are just learning about these basic homesteading principles and skills, but someday you can be so much more.

Conclusion

Homesteading is a rewarding life – no doubt about that –, but to be able to enjoy the benefits, there are certain terms to define, qualities about you that you need to discover, and skills that you need to learn.

Homesteading is also control over your health, food, home, and your life in general. And just as you would with any project, it also needs commitment, planning, a serious self-assessment, and preparation for the challenges ahead.

Many tests could come with homesteading, but the one that you need to overcome more than anything is rural loneliness. Besides this, you should also steer clear of the most common homesteader mistakes by being armed with information, designing a plan, setting the correct budget, proper land inspection, just to name a few.

Being a homesteader also means acquiring specific skills, such as composting, gardening, salvaging, food preservation and storage, milling, animal husbandry, cooking, and even the basic skill of starting a fire. And as the leader of your pack, you should also protect livestock from predators and other possible sources of harm.

Then there are the costs of the land and other homesteading equipment, machinery, and materials that should be calculated before you even make the initial offer, which is why you need to be up-to-date on the local homesteading laws and regulations. Doing so will be the only way to conduct your business without any worries.

As these things are settled, you can focus on raising your livestock and growing your produce from your garden (that you can grow better with some useful hacks). And while you are at it, you can learn some DIY crafts that can treat your boredom and even add to your income.

The beautiful end goal of homesteading is to live off the land one hundred percent. To do this, systems should be established, and you will then be able to recognize all the income-generating opportunities on your homestead.

Congratulations! You are now a well-informed homesteader.

References

(n.d.). Retrieved from https://www.youtube.com/watch?v=YJewpW3lpzc

(n.d.). Retrieved from https://youtube.com/watch?v=jO19XxEXmmQ&list=PLg8oaaTdoHz MxpqfzRHzxC-qf9Ej60dBK&index=1

(n.d.). Retrieved from https://www.youtube.com/watch?v=YvOW56x29B0

12 things to know about raising cows | Hello Homestead. (n.d.). Retrieved from https://hellohomestead.com/12-things-to-know-about-raising-cows/

Annie. (2018, June 13). Tips For Having A Large Homestead. Retrieved from https://15acrehomestead.com/large-homestead/

Annie, Cook, J., Stone, K., & Abernathy, W. T. (2019, March 31). The Ultimate Guide To Garden Soil. Retrieved from https://15acrehomestead.com/ultimate-guide-garden-soil/

Arcuri, L. (2019, May 24). What Is Homesteading? Retrieved from https://www.thespruce.com/definition-of-homesteading-3016761

Arcuri, L. (2019, December 12). How to Survive as a Beginning Homesteader. Retrieved from https://www.thespruce.com/top-tips-for-the-beginning-homesteader-3016686

Basket Making. (n.d.). Retrieved from https://www.sage-urban-homesteading.com/basket-making.html

Caitlin, Emma, Evelyn, Gibbs, L., Stephanie, Patti, ... Maria. (2020, January 9). Great Benefits of Homesteading. Retrieved from https://theelliotthomestead.com/2014/01/great-benefits-of-homesteading/

Cory, D. (2019, July 26). A Beginner's Guide to Small-Scale Homesteading. Retrieved from https://dengarden.com/gardening/Small-Town-Homesteading

Davidson, M., Davidson, M., & Alexa Weeks. (2019, December 12). 10 Most Important Homesteading Skills. Retrieved from https://homesteadsurvivalsite.com/important-homesteading-skills/

Dodrill, T. (2019, December 10). How to Find the Perfect Piece of Land for Your Homestead. Retrieved from https://homesteadsurvivalsite.com/how-to-find-land-homestead/

Garman, J., Tony, Brenda, Lana, Ana, Robin, ... Homesetading Skills. (2018, February 1). 6 Homesteading Skills to Learn First. Retrieved from https://timbercreekfarmer.com/6-homesteading-skills-to-learn-first/

Home. (n.d.). Retrieved from https://homesteadlady.com/

Homemade Candles and Candle Making. (n.d.). Retrieved from https://www.sage-urban-homesteading.com/homemade-candles.html

Homemade Paint Formula. (n.d.). Retrieved from https://www.sage-urban-homesteading.com/homemade-paint.html

Homemade Paper Making. (n.d.). Retrieved from https://www.sage-urban-homesteading.com/homemade-paper.html

Homesteading – Advantages of Self-Sufficiency on Your Own Property. (n.d.). Retrieved from http://environmentalprofessionalsnetwork.com/homesteading-advantages-of-self-sufficiency-on-your-own-property/

How Many lands You Need for a Homestead or Farm. (2018, April 13). Retrieved from https://highsierrapermaculture.com/2018/02/20/how-much-land-homestead/

Ikona, A., D., B., Brenna, Martinelli, M., Martinelli, M., Brown, M., & Marvin. (2019, October 23). Growing Trees For Profit In Your Backyard: Homesteading. Retrieved from https://homesteading.com/growing-trees-profit-backyard/

Johm, Val, Julie, Stacy, Toni, Kathryn, ... Ruth. (2018, January 30). Bringing a Barn Cat (or two) to Your Homestead. Retrieved from https://104homestead.com/barn-cat-homestead/

Kristyn, Kim, H, A., Jill, Krystyn, Amanda, ... Lorenzen, E. (2018, September 19). DIY Pallet Garden; How to make Raised Wood Pallet Garden Bed. Retrieved from https://busycreatingmemories.com/diy-pallet-garden/

Lynn, Lynn, Katie, Katie, Stacey, Stacey, ... Simple Living Country, Gal. (2019, October 6). 7 Steps to take before you begin your homesteading journey. Retrieved from https://simplelivingcountrygal.com/7-steps-to-take-before-you-begin-your-homesteading-journey/

Marie, Linda, Amanda, & L, T. (2019, November 4). What's the Big Deal GMOs? Retrieved from https://rurallivingtoday.com/homesteading-today/big-deal-gmos/

McCleod, B. R. (2015). Neo-Homesteading in the Adirondack North Country: Crafting a Durable Landscape. Neo-Homesteading in the Adirondack North Country: Crafting a Durable Landscape, 2. Retrieved from https://etd.ohiolink.edu/!etd.send_file?accession=antioch1440152751&disposition=inline

Ogden Publications, Inc. (n.d.). Guide to Urban Homesteading. Retrieved from https://www.motherearthnews.com/homesteading-and-livestock/self-reliance/guide-to-urban-homesteading-zm0z14amzrob

Ogden Publications, Inc. (n.d.). A Guide to Buying Homestead Land. Retrieved from https://www.motherearthnews.com/homesteading-and-livestock/guide-to-buying-homestead-land-zmaz82ndzgoe

Pantry Essentials: Ingredients For a Well-Stocked Homestead Kitchen. (2018, December 7). Retrieved from https://thechildshomestead.com/cooking/pantry-essentials/

Patterson, S. (2014, December 22). 3 Priceless Benefits Of Modern Homesteading. Retrieved from https://www.offthegridnews.com/how-to-2/3-priceless-benefits-of-modern-homesteading/

Pawning Through the Ages. (n.d.). Retrieved from https://www.history.com/shows/pawn-stars/articles/pawning-through-the-ages

Poindexter, J., & PoindexterJennifer, J. (2018, April 21). Raising Sheep: A Complete Guide on How to Raise Sheep at Homestead. Retrieved from https://morningchores.com/raising-sheep/

Queck-Matzie, T. (2019, March 13). Farming 101: How to Plant Corn. Retrieved from https://www.agriculture.com/crops/corn/farming-101-how-to-plant-corn

sadie423. (2019, August 13). How to Raise and Care for Goats. Retrieved from https://pethelpful.com/farm-pets/how-to-care-for-goats

Staff, M. C., & MorningChores StaffMorningChores Staff is a team of writers and editors who collaborate to create articles. If the article you are reading is authored by MorningChores Staff. (2019, August 20). 12 Things You Need to Know Before Getting Your First Ducks. Retrieved from https://morningchores.com/about-raising-ducks/

Suwak, M. (2019, October 3). 11 Fast Growing Vegetables For Your Homestead. Retrieved from https://www.primalsurvivor.net/fast-growing-vegetables/

The Editors of Encyclopaedia Britannica. (2014, May 21). Animal husbandry. Retrieved from https://www.britannica.com/science/animal-husbandry

The Ultimate Guide to Farm Dogs on the Homestead from Homestead.org. (2019, January 9). Retrieved from https://www.homestead.org/homesteading-pets/the-ultimate-guide-to-farm-friendly-dogs/

Thoma, M. (2019, September 30). How to Prepare Your Garden For Extreme Weather Conditions [37 Tips!]. Retrieved from https://tranquilurbanhomestead.com/extreme-weather/

Tipsbulletin. (2019, December 14). 70 Wonderful Uses of White Vinegar. Retrieved from https://www.tipsbulletin.com/vinegar/

Vicky, Bedford, L., Hartmann, L., & Marti. (2019, October 28). These 70 Gardening and Homesteading Hacks Will Blow You Away • New Life On A Homestead: Homesteading Blog. Retrieved from https://www.newlifeonahomestead.com/70-gardening-homesteading-hacks-will-blow-away/

Weeks, A. (2019, December 13). 12 Fun DIY Projects For Homesteaders. Retrieved from https://homesteadsurvivalsite.com/diy-projects-for-homesteaders/

Winger, J., Nichole, Nichole, Jill, Jill, Stacie, ... Christian Homefront. (2018, March 18). How to be an Apartment Homesteader • The Prairie Homestead. Retrieved from https://www.theprairiehomestead.com/2011/10/how-to-be-an-apartment-homesteader.html

https://www.youtube.com/watch?v=WLa0RRU0-G8

https://www.youtube.com/watch?v=7AUiV2zorwg

https://www.youtube.com/watch?v=E459XIluUFE

https://www.survivalsullivan.com/16-surprising-benefits-of-prepping/

https://www.youtube.com/watch?v=o9niShq9_Dg

https://www.askaprepper.com/24-prepping-items-dont-spend-money/

https://www.youtube.com/watch?v=1j5V4rJ2g5I

https://www.youtube.com/watch?v=VtcDiyj9T8k

https://www.happypreppers.com/skills.html

https://www.youtube.com/watch?v=byKqaGUiaFM

https://www.youtube.com/watch?v=Ln5qRknownw

https://www.mnn.com/lifestyle/responsible-living/stories/going-off-the-grid-why-more-people-are-choosing-to-live-life-un

https://www.youtube.com/watch?v=v8Pe_u_4q5M

https://www.youtube.com/watch?v=09IK9OvWpjw

https://www.youtube.com/watch?v=4ts15BW-6hw

https://www.youtube.com/watch?v=8-86NFf2VcE

https://thetinylife.com/common-off-grid-living-misconceptions/

https://morningchores.com/homesteading/

http://www.therealfarmhouse.com/10-steps-to-start-homesteading-on-the-cheap/

https://www.offthegridnews.com/how-to-2/9-crucial-steps-for-the-first-time-homesteader/

https://www.youtube.com/watch?v=fFHn_xoMsAs

https://www.youtube.com/watch?v=w4qcoEXYqK0

https://rurallivingtoday.com/homesteading-today/realistic-off-grid-power-sources/ https://www.treehugger.com/sustainable-product-design/generating-off-grid-power-the-four-best-ways.html

https://insteading.com/blog/off-grid-water-system/

https://www.youtube.com/watch?v=bBF72Een1D8

https://offgridworld.com/how-much-does-it-really-cost-to-go-off-grid/

https://www.youtube.com/watch?v=hJpwy9mUmho

https://www.youtube.com/watch?v=Aa74smEC0OM

https://www.youtube.com/watch?v=vDQUEuTL8tk

https://www.youtube.com/watch?v=K4wO76cqkOU

https://www.skilledsurvival.com/build-survival-medical-kit-scratch/

https://www.youtube.com/watch?v=ks00XG3n7yM

https://survivalistprepper.net/shtf-injuries-and-prevention-for-preppers/

https://unchartedsupplyco.com/blogs/news/bug-out-bag-checklist

https://www.youtube.com/watch?v=ToonXShDAFk

https://www.youtube.com/watch?v=stIjgEaES60

https://www.oldfashionedfamilies.com/6-misconceptions-about-preppers/

https://thepreppingguide.com/what-is-prepping/

https://www.shtfpreparedness.com/new-prepping-start/

https://www.askaprepper.com/24-prepping-items-dont-spend-money/

www.ingramcontent.com/pod-product-compliance
Lightning Source LLC
Chambersburg PA
CBHW050637190326
41458CB00008B/2308

* 9 7 8 1 9 5 2 5 5 9 3 7 2 *